Intelligent Freeway Transportation Systems

Robert Gordon

Intelligent Freeway Transportation Systems

Functional Design

Springer

Robert L. Gordon, P.E.
36 Stauber Drive
Plainview
NY 11803
USA
rob.gordon3@verizon.net

ISBN 978-1-4419-0732-5 e-ISBN 978-1-4419-0733-2
DOI 10.1007/978-1-4419-0733-2
Springer Dordrecht Heidelberg London New York

Library of Congress Control Number: 2009927713

Printed on acid-free paper

Springer is part of Springer Science+Business Media (www.springer.com)

Contents

Preface xi

1 Introduction 1

1.1 Purpose of Book 1
1.2 Development of ITS Design Practices 2
1.3 Summary of Contents and Organization 3
References 6

2 Cost Effective Design Processes 7

2.1 Systems Engineering 7
2.1.1 Systems Engineering Requirements for Federal Aid Projects 7
2.1.2 Systems Engineering as a Life Cycle Process 8
2.1.3 ITS Project Development 11
2.2 Goals, Objectives and Requirements 11
2.3 Evaluation Methodologies 16
References 16

3 Functional ITS Design Issues 17

3.1 Relationship of ITS Design to General Transportation Planning Principles 17
3.1.1 General Traffic Flow Relationships 18
3.1.2 Shock Waves 19
3.1.3 Classification of Congestion 19
3.1.4 Diversion for Non-Recurrent Congestion 20
3.1.5 Recurrent Congestion 23
3.2 Performance and Benefit Assessment 24
3.2.1 Performance Measures to Facilitate System Design 24
3.2.2 Performance Measures and ITS Planning 28
3.3 Alternatives for Functional Design 30
3.3.1 Design Constraints 31

3.3.2 Relationship of ITS Management Concepts to Objectives ... 32
3.4 Evaluation Concepts ... 37
3.4.1 Benefit and Cost Modeling ... 37
3.4.2 Benefit Paradigms for ITS Evaluation Models ... 38
References ... 40

4 Non-Recurrent Congestion: Improvement of Time to Clear Incidents ... 43

4.1 Description of an Incident for ITS Design Purposes ... 43
4.1.1 Effect of Incidents on Capacity ... 44
4.1.2 Secondary Accidents ... 44
4.2 Models of the Effects of Freeway Incidents ... 45
4.2.1 Frequency and Severity of Incidents ... 48
4.2.2 Data Collection for Development of Incident Model ... 50
4.3 Relationship of Reduction in Delay to Reduction in Incident Clearance Time ... 50
4.4 Interaction of Capacity Restrictions and Traffic Conditions ... 52
4.4.1 Cohort Model ... 52
4.4.2 Time Saved per Incident ... 54
4.5 Functional Requirements for Improving Incident Response and Relationship of Improvement Techniques ... 58
4.5.1 Improving Incident Detection and Verification ... 60
4.5.2 Improving Incident Response, Clearance, and Recovery Through ITS ... 67
4.6 Measuring Incident Management Effectiveness ... 72
4.6.1 Degree of Attainment for Recommended Management Functions, Operations, and Technologies ... 72
4.6.2 General Measures ... 73
4.6.3 Model for Evaluating Incident Management Effectiveness ... 73
References ... 78

5 Nonrecurrent Congestion: Incident Information to Motorists ... 81

5.1 Motorist Diversion ... 82
5.1.1 Motorist Messaging Techniques ... 82
5.1.2 Operational Diversion Policies and Strategies ... 84
5.1.3 Strategic Network Management ... 86
5.1.4 Diversion Strategies ... 90
5.1.5 Reduction in Freeway Delay Resulting from Diversion ... 94
5.1.6 Effect of Diversion on Arterial Traffic ... 95
5.1.7 Reduction in Corridor Delay Resulting from Diversion for Incidents ... 98
5.2 Design Considerations for CMS Locations ... 98

5.2.1 Basic Considerations for CMS Functional Placement 100
5.2.2 Simple Models to Assist in CMS Functional Placement 100
5.3 Quality of Motorist Information 105
5.4 ITS and Technology Applications in Emergency Evacuations 105
5.4.1 Introduction 105
5.4.2 ITS and Technology Applications 106
References 108

6 Recurrent Congestion – Information to Motorists 109

6.1 Nature of Recurrent Congestion 109
6.2 Motorist Information During Recurrent Congestion 109
6.3 Variations During Periods of Recurrent Congestion 112
6.4 Diversion During Recurrent Congestion 113
References 113

7 Ramp Metering 115

7.1 Introduction 115
7.2 Background 116
7.2.1 Early Metering Projects 116
7.2.2 Ramp Meter Installation Requirements 117
7.3 Flow Characteristics and Freeway Capacity 119
7.3.1 Flow Characteristics for Near-Capacity Conditions 119
7.3.2 Effective Capacity Improvement Through Ramp Metering 122
7.3.3 Freeway Service Improvement Through Ramp Metering 123
7.4 Ramp Metering Strategies 126
7.4.1 Overview of Metering Strategies 126
7.4.2 Pretimed Restrictive Ramp Metering 127
7.4.3 Local Traffic-Responsive Restrictive Ramp Metering 130
7.4.4 System-Wide Traffic-Responsive Restrictive Ramp Metering 134
7.4.5 Design Issues 135
7.5 Ramp Metering and the Motorist 141
7.5.1 Motorist Benefits and Disbenefits Resulting from Ramp Metering 141
7.5.2 Public Acceptance of Ramp Metering 142
7.6 Benefits Model for Ramp Metering 143
References 144

8 Communications for ITS 147

8.1 Introduction 147
8.2 Communication Standards 148

8.3 General Communication System Design Considerations 149
8.4 Communication System Technologies for Center-to-Field Communication 151
8.4.1 Wireline Based Communications 151
8.4.2 Wireless Communications 153
8.4.3 Public Network Based Communication Services 154
8.5 Example of Communications Concept Design 156
References 160

9 Transportation Management Centers 161
9.1 Transportation Management Center Functions 161
9.1.1 Support of Emergency Management Services 161
9.1.2 Provision of Information to Motorists 162
9.1.3 Operation of Ramp Meters 162
9.1.4 Operation of Service Patrols 163
9.1.5 Coordination of Traffic Signal Operation with Freeway and Corridor Requirements 163
9.1.6 Provision of Weather Information Related to Roadway Conditions 165
9.2 Information Flows Among Stakeholders 165
9.3 Implementation of Information Flows 167
9.4 Example of Transportation Management Center in Major Urban Location 167
References 173

10 Evaluation of System Design and Operation 175
10.1 Evaluation of Design Alternatives and Project Feasibility 175
10.1.1 Benefit and Cost Analysis 175
10.1.2 Alternatives Evaluation and Project Feasibility 179
10.2 Project Evaluation 180
References 181

11 Compact Disk 183
11.1 Introduction 183
11.2 System Delay per Incident 183
11.3 Relative Effectiveness of CCTV Coverage 184
11.4 Incident Management Effectiveness Potential 184
11.5 Delay Reduced on Freeway due to Queue Reduction Resulting from Diversion 184
11.6 Probability that the Motorist Encounters CMS Prior to Incident (P34) 184
11.7 Queue Storage Requirement for Ramp Meter 185
References 185

List of Symbols, Abbreviations and Acronyms 187

Appendix A Sources of Additional Information on Evaluation Models 193

Appendix B Relative Effectiveness of CCTV Coverage 195

Appendix C Example of Benefits for Incident Management 199

Appendix D Message Display Software for Southern State Parkway 203

Appendix E Washington State Fuzzy Logic Ramp Metering Algorithm 207

Appendix F Benefits Model for Motorist Assistance Patrols 211

Index 213

Preface

Functional ITS design, as described in this book, is the selection of ITS management strategies and for the field equipment deployments required to implement them. In most cases, functional design stops short of the selection of detailed technologies. Starting with the development of detailed objectives, functional design relates management strategies to project objectives, identifies alternative strategies for further consideration and evaluates these strategies. It then determines whether one or more strategies can cost effectively satisfy the objectives, and recommends the most appropriate alternative.

Although considerable effort has been expended by the Federal Highway Administration and others to develop high-level systems engineering processes, in practice, ITS designers have often used a "bottoms up" approach. Designers often select devices and device locations without a strong connection to project objectives or to methodologies that assess the feasibility of the project and the proposed design. This book provides guidance for adapting these system engineering processes to freeway ITS project functional requirements. It provides the basis for selecting the types of ITS components and the management strategies employed. A number of handbooks and other resources are available to provide guidance for the detailed selection of field equipment and operations to manage the equipment. It is assumed that the reader is familiar with the functions of ITS devices such as changeable message signs, highway advisory radio, traffic detectors and CCTV applications.

As the book emphasizes the use of fundamental transportation planning and traffic engineering principles to develop functional designs, it is assumed that the reader is somewhat familiar with this area. The book largely reflects the author's experience in adapting these principles to ITS design. For example, the book provides models to suggest appropriate locations for such ITS devices as CCTV cameras and changeable message signs, and describes methodologies for estimating the benefits of proposed functional designs. The models enable the designer to estimate the performance differences among alternatives, and estimate benefits for functional design purposes. Approximations are introduced to expedite the use of these models by practitioners. While the author has found these models to be useful, readers are encouraged to modify and enhance them to better suit their needs. The CD that accompanies the book provides worksheets that facilitate the use of some of the models. The worksheets are provided in an unprotected format to enable users to modify them as required.

I would like to express my appreciation to the book's reviewers, Mr. Richard King of Eng-Wong Taub and Associates, Dr. Lawrence Klein, Mr. Robert Reiss of Dunn Engineering Associates and Dr. Alan Stevens of the United Kingdom's Transportation Research Laboratory. They contributed many useful suggestions to organizing and clarifying the material, and to express my thoughts in a clearer and often simpler way. Mr. Emilio Sosa of the New York State Department of Transportation and Mr. Rick Knowlden of Parsons Brinckerhoff, Inc. provided descriptions and photos of the INFORM Traffic Management System. My editor at Springer, Ms. Jennifer Mirski, provided many tips to facilitate publication.

I give particular thanks to my wife, Norma, who provided support and encouragement to complete the effort.

Westhampton Beach, NY Robert Gordon

Chapter 1
Introduction

Abstract *Intelligent Freeway Transportation Systems: Functional Design* was written to provide engineers engaged in the functional design of freeway intelligent transportation systems with the background and tools to develop effective systems that are properly scaled to traffic conditions, and that use project resources efficiently. Chapter 1 describes the migration of ITS from its initial technology constraints to its current ability to support a greater variety of functions that can be integrated into a system that better satisfies stakeholders' needs.

1.1 Purpose of Book

The purpose of this book is to provide engineers engaged in the functional design of freeway ITS with the background and tools to develop effective systems that are properly scaled to traffic conditions and that use project resources efficiently. The major functions provided by freeway ITS include:

- Assisting emergency service providers in detecting and clearing incidents more rapidly. Supporting rapid response reduces the queues on the freeway with corresponding reduction in motorist delay and secondary accidents.
- Providing information on incidents and other travel conditions to motorists. Benefits accrue to motorists that choose to alter their route, trip initiation time, or travel mode. The resulting reduction in the queue on the freeway mainline also reduces the delay for motorists who do not alter their trip plan.
- Controlling access to the freeway mainline or controlling lane use. Ramp metering improves the capability of the mainline to service traffic and redistributes it on the network to reduce overall delay. Lane control signals may be used to improve traffic flow under incident conditions and improve lane allocations during normal traffic conditions.
- Managing lanes to improve freeway throughput. Examples of managed lanes include high occupancy vehicle (HOV) and high occupancy toll (HOT) lanes.

R. Gordon, *Intelligent Freeway Transportation Systems: Functional Design*,
DOI 10.1007/978-1-4419-0733-2_1,

- Managing information and controls to support a broad array of transportation needs, including transit, optimization of passenger throughput in transportation corridors, and traffic management to support responses to emergencies and evacuations.

This book provides guidance on functional design for a number of these topics.

1.2 Development of ITS Design Practices

The freeway management systems that were designed in the 1960s and 1970s were often single purpose or limited function systems. Examples include ramp metering systems, incident management in tunnels, and systems that provided limited levels of motorist information. The operational constraints on these early systems were often determined by limitations in computing and communications capability.

With improvements in computing and communications, freeway ITS were designed to provide a more comprehensive capability to manage traffic. CCTV coverage was extensively employed. Comprehensive system-wide imagery was provided by a number of systems in major metropolitan areas, and transportation management centers often operated 24 hours per day, 7 days per week. In some cases, service was extended to support high occupancy vehicle priority through management of HOV and HOT lanes and ramp meter by-passes for high occupancy vehicles.

During the late 1990s, greater emphasis was placed on the introduction of systems engineering principles into freeway ITS design. These principles are embodied in the National ITS Architecture [1]. Guidance is also provided by a California Department of Transportation and FHWA report [2]. The systems engineering principles include:

- A systematic life cycle process for establishing needs and objectives, developing a set of design alternatives, evaluating the alternatives, designing the system and installing it, providing the logistic services necessary to support the system, operating the system, and evaluating its performance for the purpose of improving future ITS designs and operations.
- Provision of a process to integrate the freeway system into the overall transportation system including support of transit, corridor operations, other regional traffic, and transportation management systems. The process emphasizes interaction with other stakeholders.
- Employing approved standards to enable interchangeability of field equipment and to support the interchange of communication among stakeholders and communication with field equipment.

Systems engineering is discussed further in Sect. 2.1.

In the late 2000s deploying freeway ITS is becoming increasingly constrained by the restricted availability of federal aid and state matching funds. Issues such as whether it is more cost effective to cover a wider geographic area with ITS treatments or to deploy ITS devices more intensively in a smaller area are important considerations for project planning and scoping.

1.3 Summary of Contents and Organization

This book describes the application of system engineering principles and cost effective design methodologies to the selection of management strategies and placement for field equipments for commonly employed ITS techniques. The relationship of traffic conditions to project scoping and preliminary design is extensively discussed.

The system engineering methodologies employ "top-down" functional design processes. Figure 1.1 illustrates the process described in this book.

The following discussion relates the book's contents to Fig. 1.1.

The first task group in the figure identifies objectives and evaluation measures. Chapter 2 describes system engineering techniques as applied to ITS and identifies goals and objectives. It relates evaluation measures to objectives, and discusses the rationale for evaluating design alternatives and project performance.

Chapter 3 prepares the groundwork for describing ITS treatments later in the book. The chapter discusses traffic conditions as they affect ITS, including traffic volume, speed, density, and capacity. Congestion and diversion are also explored. A section describing performance and benefit assessment recommends certain performance measures for assessing design alternatives. Project evaluation techniques are introduced.

The second task group in Fig. 1.1 identifies management concepts, strategies, and classes of technologies that lead to functional design. Management concepts that directly provide benefits to motorists include:

- Reduction of time to clear incidents (Chap. 4). Models of delay caused by incidents are presented, and a relationship for the delay reduced by lowering the incident clearance time is discussed. ITS techniques for improving emergency vehicle response time, such as the use of CCTV and traffic detectors,[1] are described. Procedures for maximizing the cost effectiveness of camera deployments and a trade-off of the effectiveness of traffic detector deployment and reduced incident detection time are provided. The role of the transportation management center in assisting incident responders, and a model for evaluating the effect of incident management is presented.
- Motorist information for non-recurrent congestion (Chap. 5). The role of motorist diversion and its implementation is described. The chapter discusses the capability to control the level of diversion through altering the types of messages,

[1] The terms *detectors* or *traffic detectors* are used in this book for devices that provide indications of volume, speed, or occupancy at a point on the roadway. These devices are also often called *sensors* or *traffic sensors*.

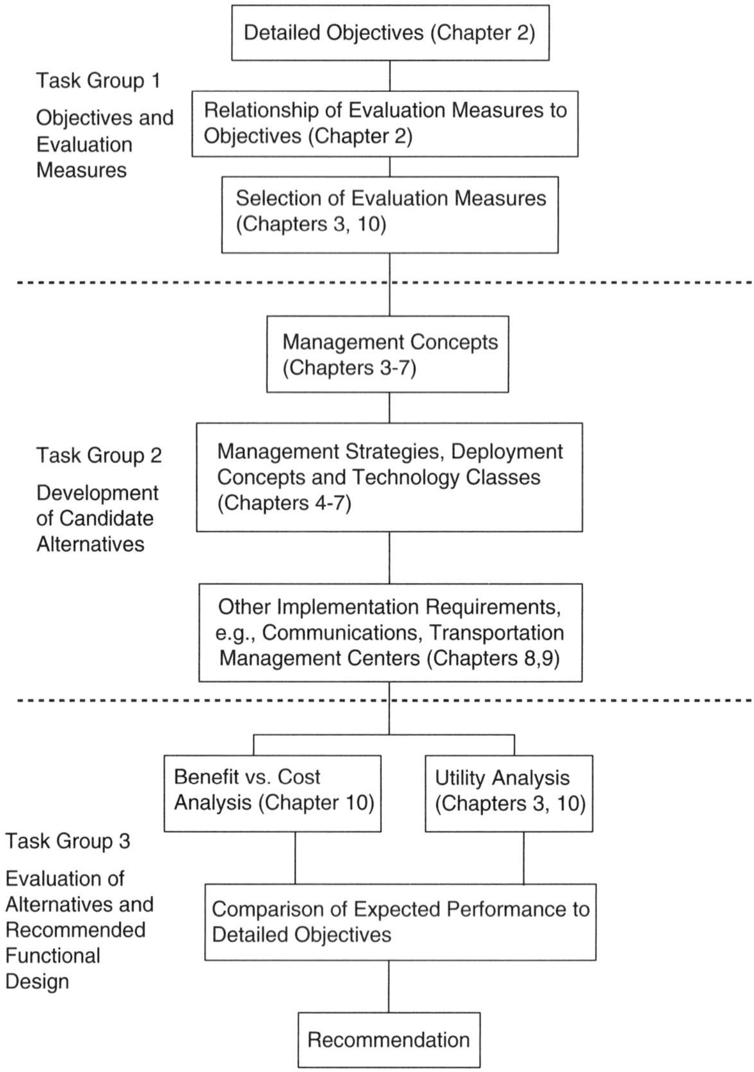

Fig. 1.1 Functional ITS design process

and presents a model to assist in obtaining the desired diversion fraction. Strategies to control the diversion fraction in real time are introduced and the effect of the diverted traffic on the alternate routes is explored. The chapter covers methods for communicating incident information to the motorist and models to optimize the placement of changeable message signs. A model to estimate the quality of information provided to the motorist is described.

- Motorist information to address recurrent congestion (Chap. 6). In addition to supplying information on non-recurrent congestion, some agencies choose to provide information on recurrent congestion. In many cases, little or no addi-

tional equipment is required to support this capability. The chapter addresses the methodologies for generating this information, and discusses how it provides a modest level of benefits that result from knowledge of traffic variations during normal capacity conditions.
- Entry ramp control. Management approaches include entry ramp closure and by-pass of entry ramp meters by high-occupancy vehicles. Chapter 7 covers ramp metering, the most commonly used entry ramp control strategy. Guidance is provided to determine whether ramps are suitable for metering. The ability of metering to improve merge characteristics and increase capacity is covered, as it the use of metering that helps control access to the freeway's mainline. Basic strategies include restrictive and non-restrictive metering and their effects on ramp queues and diversion. Sub-strategies include pretimed and traffic-responsive metering as well as isolated and system-wide metering. Design issues, impacts of traffic diversion and accommodations that may be necessary to achieve motorist acceptance of ramp metering are described, and a model for obtaining benefits is provided.

While the book discusses a number of key freeway management concepts in detail, a number of additional strategies are available. Neudorff [3] discusses many of these concepts.

Functional design alternatives may differ in the following ways:

- Different selections of the management strategies identified above and different choices of sub-strategies (e.g. type of ramp metering and ramps to be metered, and motorist messaging strategies.
- Intensity of implementation (e.g. average number of CCTV cameras, detector stations, and changeable message signs per mile of roadway).
- Types of services provided by the transportation management center such as hours of operation, and interaction with other agencies such as transit and emergency responders.

An overview of freeway ITS communications is provided by Chap. 8. The use of the Open Systems Interconnect (OSI) model and the National Transportation Communications for ITS Protocol (NTCIP) standards are described. The chapter discusses the features of wireline and wireless technologies, along with the issues of implementing the technologies by utilizing communication service providers or ownership by the operating agency. An example of a communications concept design for a small system illustrates the selection of technology alternatives based on communication reliability and cost.

Chapter 9 discusses the support provided for the following functions by transportation management centers:

- Support of emergency management services.
- Provision of information to motorists.
- Operation of ramp meters.
- Operation of service patrols.
- Coordination of traffic signals with freeway and corridor operations.

- Provision of weather related information to motorists.

The chapter describes methods for establishing information pathways among stakeholders, and provides an example of how the flows that are required by a regional ITS architecture may be implemented. The functions of the INFORM Traffic Management Center, a TMC in a major metropolitan area, illustrate the application of the concepts in the chapter.

The third task group in Fig. 1.1 leads to a recommendation for functional design. Benefit vs. cost analysis is, in many cases, the key to recommending further continuation of the project, and identifying the design alternative to be implemented. Chapter 10 describes how evaluation models may be used to estimate benefits and transform them into a monetary format suitable for benefit vs. cost analysis. The discussion includes an example of how the results of the analysis may be effectively presented.

Utility analysis is discussed in Chaps. 3 and 10. It may be used to estimate benefits that are difficult to estimate in terms of dollars. It is also useful for combining monetary and non-monetary benefits. Chapter 10 also provides examples of how benefit and cost analysis may be used to recommend a design alternative for implementation.

A compact disk containing worksheets to support several methodologies discussed in the book is provided. The use of these worksheets is described in the relevant chapters. Chapter 11 identifies these worksheets and provides a brief description of their application.

References

1. National ITS Architecture V6.1. U.S. Department of Transportation. www.iteris.com/itsarch/. Accessed 27 January 2009
2. Systems Engineering Guidebook for ITS, Version 2.0 (2007), California Department of Transportation and Federal Highway Administration
3. Neudorff LG et al (2003) Freeway management and operations handbook. Report FHWA-OP-04-003, Federal Highway Administration, Washington, DC

Chapter 2
Cost Effective Design Processes

Abstract Within the context of the design and life cycle operation of a freeway ITS project, this chapter discusses systems engineering requirements for federal aid projects, and the adaptation of these requirements to the design and project development process used by many agencies. It describes objectives for freeway management systems and their relationship to project benefits as well as methodologies for the evaluation of design alternatives, project feasibility, and project life cycle evaluation.

2.1 Systems Engineering

2.1.1 Systems Engineering Requirements for Federal Aid Projects

Part 940 of Title 23 of the Code of Federal Regulations (23CFR940) describes the requirements for receiving federal aid. Key provisions include the availability of a Regional ITS Architecture and project-based systems engineering analysis. The regulation states that the regional ITS architecture comprises a regional framework for insuring institutional agreement and technical integration for the implementation of ITS projects or groups of projects. To qualify for federal aid, ITS projects must conform to the National ITS Architecture [1], a framework on which the regional ITS architecture is based. The regional ITS architecture includes:

- Development of operational concepts and agreements among the participating regional agencies and stakeholders.
- System functional requirements.
- Interface requirements and information exchanges with planned and existing systems.

R. Gordon, *Intelligent Freeway Transportation Systems: Functional Design*,
DOI 10.1007/978-1-4419-0733-2_2,

- Identification of ITS standards.
- Sequence of projects required for implementation.

A project-level ITS architecture that is coordinated with the regional ITS architecture is also required by 23CFR940. The systems engineering analysis for the project level architecture includes, as a minimum:

- Identification of the portions of the regional ITS architecture being implemented, and identification of cognizant agencies and responsibilities.
- Definition of requirements.
- Analysis of alternative system configurations and technologies.
- Procurement options.
- Identification of applicable ITS standards and testing procedures.
- Procedures and resources necessary for operation and management of the system.

2.1.2 Systems Engineering as a Life Cycle Process

The concept of systems engineering developed significantly in the 1960s with the advent of large military and space systems. It represents the amalgamation of a number of engineering disciplines together with economics, human factors, goal setting and evaluation techniques [2].

The International Council on Systems Engineering (INCOSE) defines Systems Engineering as "an interdisciplinary approach and means to enable the realization of successful systems. It focuses on defining customer needs and required functionality early in the development cycle, documenting requirements, then proceeding with design synthesis and system validation while considering the complete problem…

Systems Engineering integrates all the disciplines and specialty groups into a team effort forming a structured development process that proceeds from concept to production to operation. Systems Engineering considers both the business and the technical needs of all customers with the goal of providing a quality product that meets the user needs" [3].

System engineering practices differ considerably depending on the application. Advanced military systems, for example, emphasize, development of technology, while ITS design decisions principally utilize existing components and adaptations of existing software to minimize costs. The Systems Engineering Guidebook for ITS [4] provides project guidance for implementation by applying the Vee technical development model (Fig. 2.1). The Guidebook indicates that the items in the Vee diagram ranging from Concept Exploration through System Requirements are influenced by the regional ITS architecture.

A key element in the Vee diagram is the Concept of Operations (CONOPS). The objectives of the CONOPS are to document the total environment and use of the system from the point of view of the stakeholders. The Guidebook provides the template shown in Table 2.1 to assist in CONOPS development. Reference [5] provides additional guidance for CONOPS details.

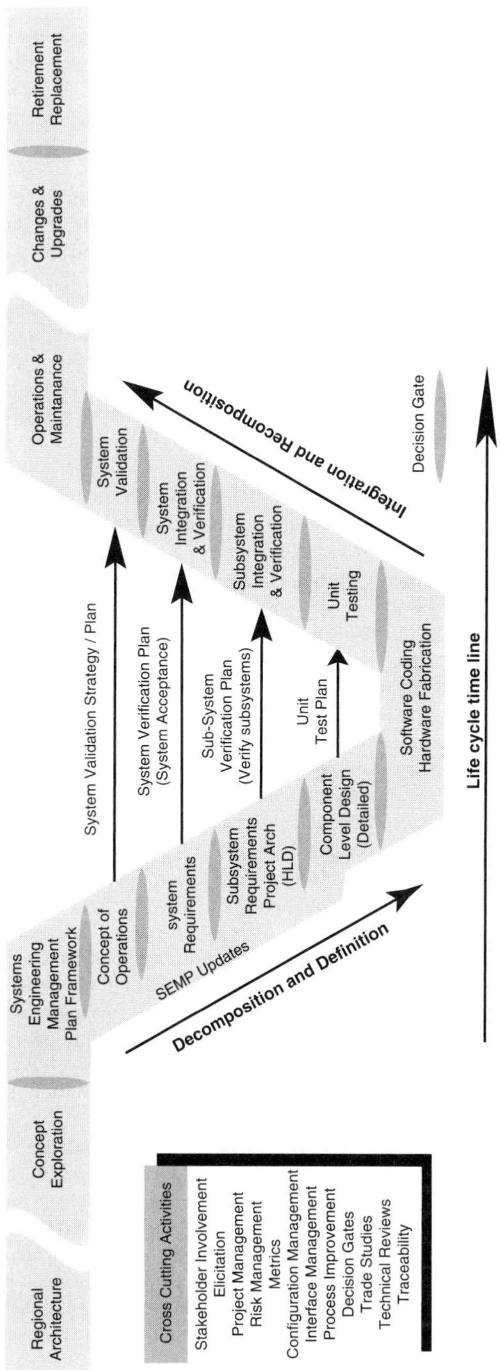

Fig. 2.1 Vee technical development model [5]

Table 2.1 Template for concept of operations

Section	Contents
Title Page	The title page should follow the Transportation Agency procedures or style guide. At a minimum, it should contain the following information: • CONCEPT OF OPERATIONS FOR THE [insert name of project] AND [insert name of transportation agency] • Contract number • Date that the document was formally approved • The organization responsible for preparing the document • Internal document control number, if available • Revision version and date issued
1.0 Purpose of document	This section is a brief statement of the purpose of this document. It is a description and rationale of the expected operations of the system under development. It is a vehicle for stakeholder discussion and consensus to ensure that the system that is built is operationally feasible. This will briefly describe contents, intention, and audience. One or two paragraphs will suffice.
2.0 Scope of project	This short section gives a brief overview of the system to be built. It includes its purpose and a high-level description. It describes what area will be covered and which agencies will be involved, either directly or through interfaces. One or two paragraphs will suffice.
3.0 Referenced documents	This optional section is a place to list any supporting documentation used and other resources that are useful in understanding the operations of the system. This could include any documentation of current operations and any strategic plans that drive the goals of the system under development.
4.0 Background	Here is a brief description of the current system or situation, how it is used currently, and its drawbacks and limitations. This leads into the reasons for the proposed development and the general approach to improving the system. This is followed by a discussion of the nature of the planned changes and a justification for them.
5.0 Concept for the proposed system	This section describes the concept exploration. It starts with a list and description of the alternative concepts examined. The evaluation and assessment of each alternative follows. This leads into the justification for the selected approach. The operational concept for that selected approach is described here. This is not a design, but a high-level, conceptual, operational description. It uses only as much detail as needed to be able to develop meaningful scenarios. In particular, if alternative approaches differ in terms of which agency does what, that will need to be resolved and described. An example would be the question of whether or not a regional signal system will have centralized control.
6.0 User-oriented operational description	This section focuses on how the goals and objectives are accomplished currently. Specifically, it describes strategies, tactics, policies, and constraints. This is where the stakeholders are described. It includes who users are and what the users do. Specifically, it covers when, and in what order, operations take place, personnel capabilities, organizational structures, personnel & inter-agency interactions, and types of activities. This may also include operational process models in terms of sequence and interrelationships.

2.1.3 ITS Project Development

While the Vee model provides a basis for project development, many of the agencies that acquire and operate ITS traditionally use procedures that are largely based on those used for highway construction. For example, the New York State Department of Transportation (NYSDOT) develops projects, including ITS projects through a process described in the New York State Department of Transportation Project Development Manual [6]. This manual describes the following stages of project development:

- Initial Project Proposal – A preliminary description of the problem, project objectives, schedule and cost estimate. For ITS projects this is usually a brief document.
- Project Scoping Stage – Identifies conditions, needs, objectives, design criteria, feasible alternatives, and cost. For ITS projects the stage usually results in a functional design for a recommended alternative and a plan for general deployment of major devices such as CMS.
- Design Stage – This stage is broken into a number of phases. A design document is prepared to summarize the results of the preliminary design phases, and the final design phases result in a set of plans, specifications, and estimates (PS&E) for the project.

In order to adapt the NYSDOT project development process to ITS needs and facilitate the incorporation of federal system engineering requirements, Appendix 6 (Intelligent Transportation Systems Scoping Guidance) was incorporated into the Project Development Manual. Figure 2.2 shows an overview of the scoping guidance provided in that appendix.

Table 2.2 shows the relationship of the major NYSDOT project stages and operations functions to the Vee diagram. The identification numbers in the scoping column correspond to the processes in Fig. 2.2.

2.2 Goals, Objectives and Requirements

Systems for various technical fields are often developed to satisfy a specific set of objectives or requirements. For example, the acceptance of a newly designed airplane by an airline is often based on its performance with respect to a set of previously agreed specifications such as the attainment of a speed versus altitude envelope.

ITS objectives or requirements are usually formulated from a more general set of goals established by stakeholders. The National ITS Architecture [1] defines goals in a broad manner for all ITS applications. These goals are:

- Improve the safety of the nation's surface transportation system.
- Increase the operational efficiency and capacity of the surface transportation system.
- Reduce energy consumption and environmental costs.

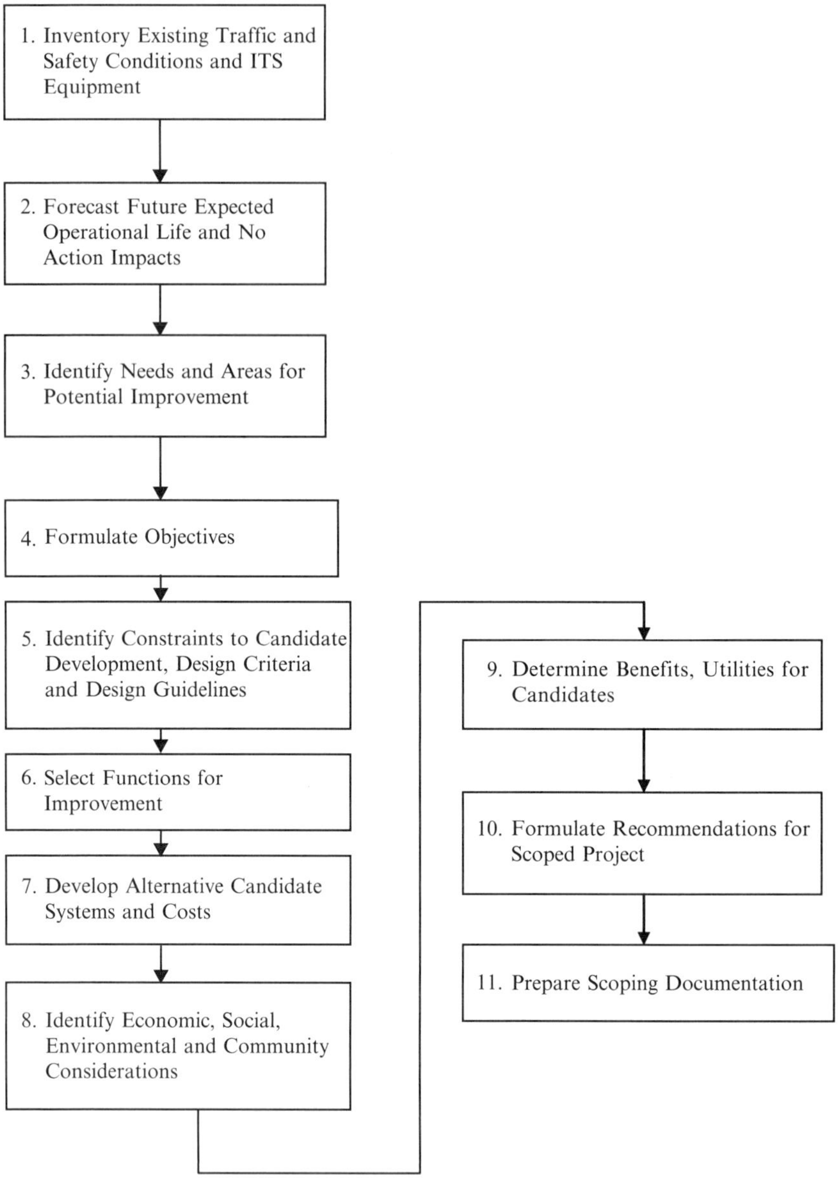

Fig. 2.2 ITS project scoping process overview [6]

- Enhance present and future economic productivity of individuals, organizations, and the economy as a whole.
- Enhance the personal mobility and the convenience and comfort of the surface transportation system.
- Create an environment in which the development and deployment of ITS can flourish.

Table 2.2 Relationship of NYSDOT project development and operation to systems engineering life cycle [6]

FHWA VEE diagram process	NYSDOT engineering project development and operations functions				
	Scoping	Preliminary design	Detailed design	Const-ruction	Operations and maintenance
Concept of operations	Traffic conditions and safety inventory (1)				
	Stakeholders, participating agencies (3)				
	Objectives (4)				
	Measures of effectiveness (4)				
High level requirements	Future service life (2)				
	Constraints (5)				
	Selection of functions for improvement (6)				
	Development of alternatives (7)				
	Other considerations (8)				
	Benefits (9)				
	Recommendations for project development (10)				
Detailed requirements		√			
High level design		√			
Detailed design			√		
Implementation				√	
Integration and test				√	
Subsystem verification				√	
System verification				√	
Operations & maintenance					√
Assessment					Performance measures from Step 4

For the purpose of scoping advanced traffic management systems, these general goals may be focused into general objectives for a project. An example of a candidate set of general objectives for both freeway and surface street systems is shown in Table 2.3 [6]. The objectives for a project should be selected by the stakeholders, and related to the regional ITS architecture. To be effectively employed, objectives should be measurable. Thus the table includes a set of possible evaluation measures. A freeway management project will typically employ a subset of these objectives.

Table 2.3 General candidate objectives [6]

Objective	Possible evaluation measure
1. Reduce congestion and improve travel time	
a. Recurrent congestion/travel time – significant section	Vehicle hours per year saved, person hours per year saved.
b. Recurrent congestion – spot	Vehicle hours per year saved, person hours per year saved.
c. Nonrecurrent congestion – significant section	Vehicle hours per year saved, person hours per year saved.
d. Nonrecurrent congestion – spot	Vehicle hours per year saved, person hours per year saved.
2. Reduce accident rate	
a. Over significant section	Reduction in accidents per year.
b. Spot	Reduction in accidents per year.
3. Reduce emissions and fuel consumption	Reduction in grams of HC, NOX, CO per year.
4. Serve as a corridor link in a wider area highway system	Provide capability for this roadway to offer meaningful diversion opportunity for incidents on another freeway that carries interregional traffic.
5. Serve as a diversion route in local corridor	Provide capability for this roadway to offer meaningful diversion opportunity for incidents on another freeway or surface street that carries local or intraregional traffic.
6. Special traffic management functions	
a. Traffic monitoring for major roadway reconstruction	Provision of capability to assist in management of traffic during construction.
b. Traffic monitoring for minor roadway construction	Provision of capability to assist in management of traffic during construction.
c. High occupancy vehicles	Provision of ITS support for HOV Operations.
d. Signal preemption for railroad or emergency vehicles	Reduction in accidents per year.
e. Priority for transit	Reduction in total traveler hours per year. Reduction in total grams of HC, NOX, CO per year.
f. Traffic information on roadway construction	Increase in motorists notified.
g. Motorist information on travel conditions, parking, special events, roadway weather	Increase in motorists notified.
h. Pedestrian and bicycle movement	Reduction in total traveler hours per year. Reduction in accidents per year.
i. Mobility and safety for the disabled	Annual volume of trips for the disabled. Average travel time for trip for disabled. Number of accidents per year for disabled.

In establishing general project objectives, stakeholders should be aware of the relative levels of benefits that ITS treatments are likely to provide. For example, Fig. 2.3 shows the relative benefits achieved by several ITS treatments for a typical

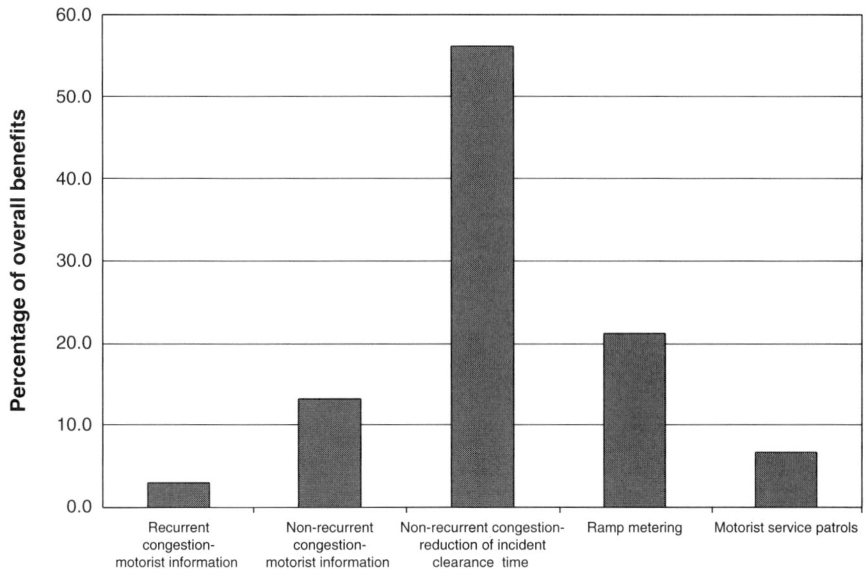

Fig. 2.3 Typical relative improvement for several ITS measures

freeway whose peak period volume-to-capacity ratio is generally between 0.93 and 0.96. It is seen that objectives that result in the reduction of non-recurrent congestion through improvement of incident clearance time are likely to provide the greatest benefits.

To assist in evaluating design alternatives after selection of the subset of general objectives used for the project, it is useful to further tailor this subset to the project, and provide additional detail. Accordingly specific evaluation measures are selected and objectives are quantified to the extent possible. Exploiting quantitative benefit values to express objectives is important because it provides both system designers as well as decision makers with a means to identify expectations, estimate the scale and scope of the project and determine its utility relative to other candidate projects. It also provides a basis for assessing the effectiveness of the design. The values should be sufficiently high to make a meaningful impact, but must also be achievable by technologies and resources available for the project. ITS benefits models (Section 3.4.1) and the Research and Innovative Technology Administration (RITA) ITS benefits data base [7] provide guidance for the selection of values for quantified objectives. We use the term *detailed project objectives* instead of *requirements* in order to provide some flexibility for stakeholder review and reassessment based on project evaluation. Other features of detailed objectives include the following [8]:

- Detailed objective should stand alone, that is, it should not require the reader to look at additional text.
- Detailed objective should be open to only one interpretation.
- Detailed objective should be verifiable through inspection, analysis, or test.

2.3 Evaluation Methodologies

Evaluation methodologies may be employed for the following major functions in the life cycle process:

- Design evaluations to select the most appropriate design alternatives and to estimate the expected performance of these alternatives relative to the detailed objectives. These *prospective* evaluations (Section 3.2) are most easily performed using the types of models discussed in Section 3.4.1.
- Project performance evaluations are *retrospective* evaluations (Section 3.2) performed to determine actual system performance and cost relative to expectations. They may be executed by field measurements, measurements made by system detectors, accident statistics, and actual cost information. In some cases simulation is used. These evaluations not only evaluate the system design and operation but also provide feedback for improving the operation of the current project, and to provide guidance for the functional design of future projects.

References

1. National ITS Architecture V6.1, www.iteris.com/itsarch/. Accessed 27 January 2009
2. Gordon RL (2003) Systems engineering processes for developing traffic signal systems, NCHRP Synthesis 307. Transportation Research Board
3. International Council on Systems Engineering (INCOSE), www.incose.org. Accessed 28 October 2008
4. Systems engineering guidebook for ITS, version 2.0 (2007), California Department of Transportation and Federal Highway Administration
5. Configuration management for transportation management systems transportation systems – establishing and maintaining systems integrity, http://ops.fhwa.dot.gov/freewaymgmt/publications/cm/presentation/index.htm. Accessed January 9 2009
6. New York State Department of Transportation project development manual (2004) New York State Department of Transportation
7. Intelligent transportation systems benefits database, Research and Innovative Technology Administration, http://www.benefitcost.its.dot.gov/its/benecost.nsf/BenefitsHome. Accessed 31 October 2009
8. Developing functional requirements for ITS projects (2002), Mitretek Systems

Chapter 3
Functional ITS Design Issues

Abstract Chapter 3 discusses a number of topics that form the conceptual framework and the building blocks for developing ITS designs. Topics include:

- *Relationship of ITS design approaches to general transportation planning principles.* The impacts of volume, speed, density, capacity issues, recurrent and non-recurrent congestion are introduced and discussed. Insights into the decisions that influence how motorists make diversion choices to alternate routes are presented.
- *Performance and benefit assessment.* Performance measures for evaluation of ITS design alternatives are needed. Therefore marginal analysis and multi-attribute utility analysis are described and illustrated as examples of alternatives to the use of traditional cost benefit analysis.
- *Alternatives for functional analysis.* The need for an alternatives analysis is presented along with a discussion of the constraints that typically affect alternative options. A matrix identifying the relationship of ITS management concepts to project objectives is presented in order to facilitate consideration of the appropriate ITS treatment to achieve these objectives.

3.1 Relationship of ITS Design to General Transportation Planning Principles

ITS functions and designs must be consistent with general traffic engineering and transportation planning principles. The following sections describe a number of these principles that affect ITS concept design.

R. Gordon, *Intelligent Freeway Transportation Systems: Functional Design*,
DOI 10.1007/978-1-4419-0733-2_3,

3.1.1 General Traffic Flow Relationships

3.1.1.1 Relationship of Volume, Speed, and Traffic Density

This relationship is provided in various texts (for example see May [1]) as

$$q = k \bullet u \tag{3.1}$$

where
q = volume (vehicles per hour per lane).
k = density or concentration (vehicles per mile per lane).
u = space-mean-speed (miles per hour).

Space-mean-speed is the average speed of vehicles measured over a short distance, for example by computing the average vehicle speed from the vehicle travel times measured over the distance. Probe traffic detection techniques (see Sect. 4.5.1.3) measure space-mean-speed. Time-mean-speed (u_T) is the average speed of vehicles measured at a point, and may be measured by point detectors (Sect. 4.5.1.3). These quantities are statistically related [2]. A typical empirical relationship is [3]:

$$u = 1.026 \bullet u_T - 1.89 \tag{3.2}$$

where u and u_T are in miles per hour. Time-mean- speed exceeds space-mean-speed with the largest divergence at low speeds.

3.1.1.2 Capacity

Capacity is defined as "the maximum sustained 15-min flow rate, expressed in passenger cars per hour per lane, that can be accommodated by a uniform freeway segment under prevailing traffic and roadway conditions in one direction of flow [4]." Reference [4] presents the suggested speed-flow relationship shown in Fig. 3.1.

The figure shows three flow regimes for basic freeway segments. In the undersaturated regime, as traffic volume increases, speed decreases slightly until the capacity is reached (2,250–2,400 passenger cars per hour per lane (pc/h/ln) depending on the free flow speed). Chapter 7 discusses the possibility of increasing this capacity by smoothing ramp entry flows. The queue discharge regime represents the flow resulting from a bottleneck (when freeway demand exceeds capacity). Flow rates of from 2,000 to 2,300 pc/h/ln are usually experienced in this regime for basic freeway sections.

Oversaturated flow is influenced by the effects of a downstream bottleneck (i.e. by flow into existing congested conditions). Low speeds and high values of density characterize the congestion in this regime.

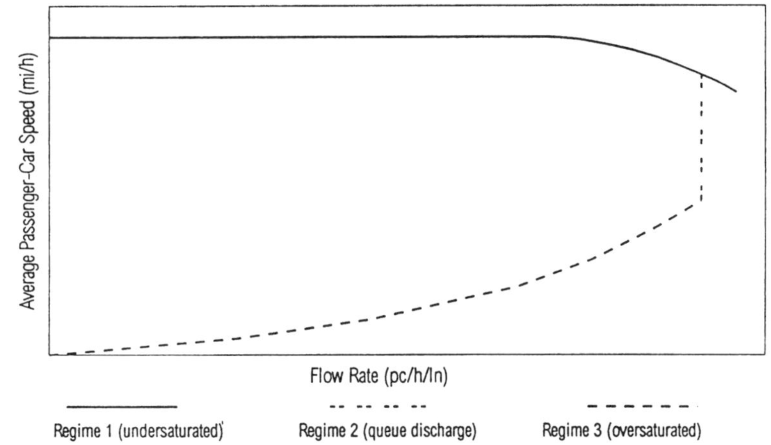

Fig. 3.1 Speed vs. flow rate. *Source*: Highway Capacity Manual 2000. Copyright, National Academy of Sciences, Washington, D.C., Exhibit 13-4, pp. 13-5. Reproduced with permission of the Transportation Research Board. Regime references added by author

3.1.2 Shock Waves

Shock waves are defined as boundary conditions in the time-space domain that demark a discontinuity in flow-density conditions [1]. The motorist may experience significant speed changes at shock wave boundaries. The propagation speeds of shock waves often determine the time it takes to detect an incident with ITS devices. Section 4.5.1.3 discusses these issues.

3.1.3 Classification of Congestion

Congestion is travel time or delay in excess of that normally incurred under light or free-flow travel conditions. Unacceptable congestion is travel time or delay in excess of an agreed-upon norm. The agreed-upon norm may vary by type of transportation facility, travel mode, geographic location, and time of day, and should be derived taking into account the expectations for each portion of the transportation system as influenced by community input and technical considerations [5]. Many transportation agencies consider a "D" level of service [4] as acceptable and use it as the baseline for defining unacceptable delay.

Congestion is often classified as either recurrent or non-recurrent. The type of congestion depends on whether the capacity or the demand factor is out of balance.

- *Recurrent* congestion occurs when demand increases beyond the available capacity. It is usually associated with the morning and afternoon work commutes, when demand reaches such a level that the freeway is overwhelmed and traffic flow deteriorates to unstable stop-and-go conditions [5]. The role of ITS in reducing recurrent congestion is discussed in Sect. 3.1.5 and in Chap. 6.

- *Non-recurrent* congestion results from a temporary decrease in capacity while the demand remains unchanged. This kind of congestion usually results when the freeway capacity becomes temporarily restricted. A stopped vehicle, for example, can take a lane out of service; however, the same number of vehicles require passage. Speed and volume drop until the lane is reopened, and the freeway returns to full capacity. Capacity can also be decreased by weather events and events near the travelway (i.e., "rubber necking" also known as "gawking"), leading to non-recurrent congestion and reduced reliability of the entire transportation system [1]. ITS techniques to reduce the effect of non-recurrent congestion include reducing the time to clear an incident (Chap. 4) and motorist diversion (Sect. 3.1.4 and Chap. 5).

3.1.4 Diversion for Non-Recurrent Congestion

3.1.4.1 Diversion Decisions by Motorists

The motorist normally selects a route based on the comparative utility (travel time and other factors) of alternative route choices. Traffic assignment models estimate the choices by user class. Other factors being equal, the freeway route is selected by most users under normal conditions.

When an incident has happened on the freeway and motorists are aware of a problem, some percentage of motorists using the freeway will choose to divert to an alternate route.

Individual diversion decisions are made on the basis of perceived time saved or other factors such as trip reliability based on information available to the motorists from various sources, along with their perception of the traffic environment. Studies show that the propensity for a motorist to divert depends on the "strength" of the message. In Table 3.1 Peeta et al. show an example of classifying increasing message strength [6].

Table 3.1 Example of increasing message strength

Message type	Message content
1	Occurrence of accident only
2	Location of the accident only
3	Expected delay only
4	The best detour strategy only
5	Location of the accident and the best detour strategy
6	Location of the accident and the expected delay
7	Expected delay and the best detour strategy
8	Location of the accident, expected delay and the best detour strategy

Source: ref. [6]. Reproduced with permission of the Transportation Research Board

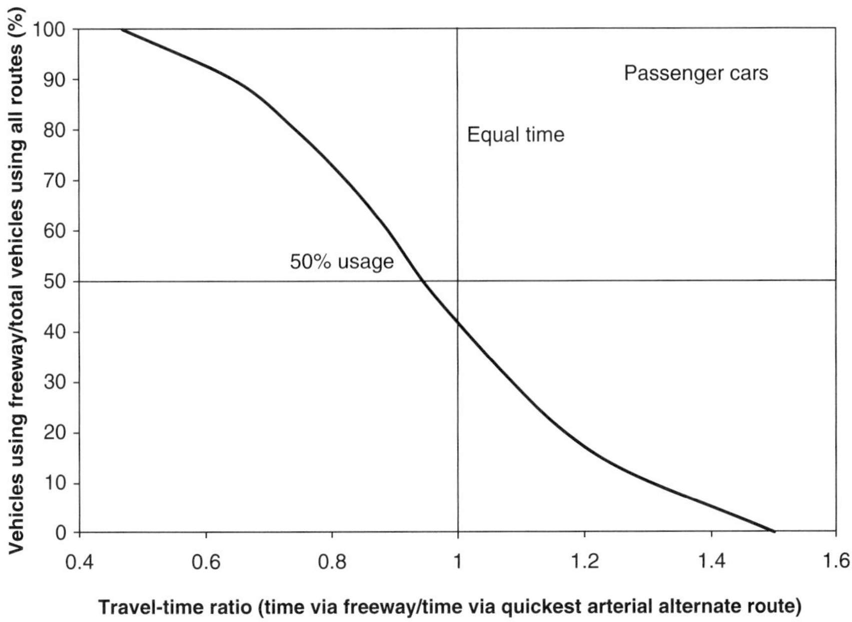

Fig. 3.2 Bureau of Public Roads Diversion Curve [7]

Operating agencies may consider the degree of diversion desired in formulating message displays (see Chap. 5).

Estimates and models have been developed that show the probability of diversion based on the perception of time saved. An early example of an assignment curve is shown in Fig. 3.2.

Using data from Ullman et al. [8], Fig. 3.3 depicts the diversion percentage from a freeway to a surface street for motorists receiving a message of strength 8 in Table 3.1 for a particular set of circumstances. The slope of the curve for time savings of under 20 min is approximately 6% diversion per minute of time saved.

Additional discussion of diversion is provided in Chap. 5.

3.1.4.2 System-Wide Impacts of Diversion

If traffic information concerning an incident is provided before the start of the trip or early in the trip, motorists may choose to:

- Select an alternate route
- Select an alternative time for the trip
- Select a different mode

For most incidents, information reaching the motorist during the freeway trip will provide the motivation to divert, most often, to a surface street. The previous

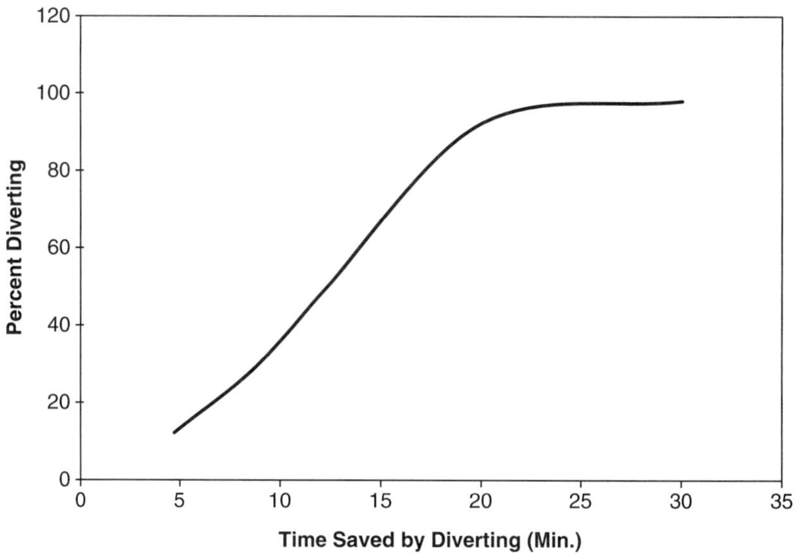

Fig. 3.3 Diversion from freeway to surface street

section discussed likely motorist responses to this information. This section discusses system considerations in providing motorist information.

From a transportation system perspective, the objective of diversion is to minimize delay in the corridor and/or maximize mobility (throughput) in the corridor. While diversion, if properly implemented, will reduce delay to the vehicles diverting and reduce delay to the freeway vehicles that do not divert, it will cause additional delay to the vehicles that normally use the diversion route. Some of these vehicles may, in turn, divert to a roadway with a lower classification rating (e.g. from a principal arterial street to a minor arterial street) generally increasing the delay to these vehicles.

Most traffic management centers manage the provision of motorist information by means of operating procedures. These procedures reflect the policies of the agency and contain message structure guidelines as well as the circumstances that require messages. Dudek [9] provides a comprehensive discussion of these factors and a number of examples of policy statements for various aspects of message development and changeable message sign operation. The example policy for diversion using changeable message signs (CMS) is:

> When incidents occur that do not require the full closure of the roadway and it is desirable to divert traffic from the freeway, CMS messages shall not divert motorists to specific alternative routes unless positive guidance is available along the alternative route in the form of a) guide signs and/or trailblazers to the major destination, or b) law enforcement or traffic control personnel positioned at critical locations along the alternative route to control and guide traffic.

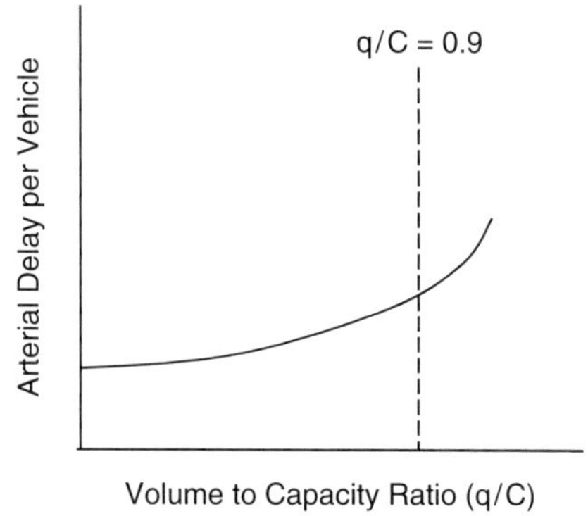

Fig. 3.4 Relationship of arterial delay to volume- to-capacity ratio

Furthermore, both of the following conditions must be met:

- The CMS operator has current and continuously-updated knowledge of the traffic conditions on the alternative route; and
- The alternative route will result in a significant savings in time for the diverted motorists.

This policy is designed to apply to explicit diversion messages (e.g. message types 4, 5, 7, and 8 in Table 3.1). Although the other message types do not explicitly recommend diversion, they do, in fact, result in diversion, and the TMC operating procedures should reflect this.

Improvements in delay to non-diverting freeway traffic may be significant, even for relatively low levels of diversion. This is discussed further in Chap. 5.

Even low levels of freeway diversion may cause significant increases in the volume-to-capacity ratio on surface street alternates. When this volume-to-capacity ratio exceeds 0.9, the delay increases exponentially as shown in Fig. 3.4. Operating procedures must ensure that arterial delay does not become excessive.

3.1.5 Recurrent Congestion

Recurrent congestion is characterized by congestion that repeats on a day-to-day basis at the same locations, usually during peak traffic periods. Most motorists are familiar with traffic conditions on the freeway and on its alternates during these periods. Under these conditions, traffic distribution on the network is determined by Wardrop's Principles [10]. These principles are summarized as follows:

- Wardrops's First Principle (user equilibrium) indicates that the trip costs on all routes actually used are less than those that would occur if a vehicle used any unused route. Thus users choose the route that minimizes their own travel time.

- Wardrop's Second Principle (system equilibrium) indicates that cost of using the system is at a minimum. Users distribute themselves so that the average travel time on each origin–destination pair is equal for all users.

The costs include travel time and fuel consumption as well as other costs that might be perceived by drivers. According to these principles, there is nothing to be gained by selecting an alternate route, as such a selection will lead to a higher cost. This case is markedly different from the case of non-recurrent congestion where the motorist has incomplete knowledge of network conditions.

With these principles in mind, most ITS operators do not provide information on changeable message signs to address recurrent congestion. As discussed in Sects. 6.3 and 6.4, opportunities to improve travel time by providing recurrent congestion traffic information may be present under some conditions. Thus some agencies do provide this information for reasons discussed in Chap. 6.

Measures may be considered to increase traveler throughput under recurrent congestion by biasing the delays or motorists' cost in the network. Measures such as restrictive ramp metering and lane or roadway use controls provide motorists with an altered set of link travel times, and in some cases out of pocket cost. The application of Wardrop's principles to this new configuration will alter the travel pattern.

3.2 Performance and Benefit Assessment

Two types of evaluations may be conducted:

- *Prospective evaluations*: During the early part of the concept design for the project scoping and the design process that follows, prospective evaluations assist in deciding whether the system should be implemented, and provide the basis for the evaluation of design alternatives. Title 23 Section 940.11 of the Code of Federal Regulations (CFR) requires the evaluation of design alternatives for a project to be considered for federal aid (see Chap. 2).
- *Retrospective evaluations*: Performed after system installation and operation, retrospective evaluations provide the basis for developing "lessons learned" and for the future improvement of system design and operation.

Reference [11] provides an in-depth discussion of the issues concerning the selection of performance measures for freeway applications. Table 3.2 [12] provides a set of recommended measures for highway applications:

3.2.1 Performance Measures to Facilitate System Design

This book is primarily concerned with the development and evaluation of design alternatives and therefore primarily focuses on economic benefits and costs for design and prospective evaluation purposes.

Table 3.2 Recommended performance measures for highways

Outcomes (operational) performance measures
Quantity of travel (users' perspectives)
Person-miles traveled
Truck-miles traveled
Vehicle miles traveled
Persons moved
Trucks moved
Vehicles moved
Quality of travel (users' perspectives)
Average speed weighted by person-miles traveled
Average door-to-door travel time
Travel time predictability
Travel time reliability (of trips that arrive in acceptable time)
Average delay (total, recurring, and incident-based)
Level of Service (LOS)
Utilization of the system (agency's perspective)
Percent of system heavily congested (LOS E or F)
Density (passenger cars per hour per lane)
Percentage of travel heavily congested
Volume-to-capacity ratio
Queuing (frequency and length)
Percent of miles operating in desired speed range
Vehicle occupancy (persons per vehicle)
Duration of congestion (lane-mile-hours at LOS E or F)
Safety – Incident rate by severity (e.g., fatal, injury) and type (e.g., crash, weather)
Incidents
Incident induced delay
Evacuation clearance time
Outputs (agency performance)
Incident response time by type of incident
Toll revenue
Bridge condition
Pavement condition
Percent of ITS equipment operational

Source: From ref. [12]

Many of the measures in Table 3.2 represent different ways of expressing the same economic benefit to be achieved by ITS. In order to be useful for evaluating alternatives and developing recommendations for design implementation, it is necessary to select a subset of these measures that facilitates the evaluation of the following categories:

- Economic benefits. For evaluation purposes, delay reduction, safety, and fuel consumption will be considered in this category.
- Environmental benefits (emissions).
- Mobility.
- Public satisfaction with ITS treatments.

To provide a basis for evaluating alternatives and developing recommendations, it is necessary that the measures selected include all of the benefits in a category, but do not include each benefit more than once. For example, the measures selected for economic evaluations should include an estimate of all of the economic benefits but should not include the same effect (e.g. travel time and delay) more than once. Table 3.3 suggests a set of measures that may be used to evaluate system designs.

Measures intended for design evaluations are often based on analytic or simulation processes, whereas ongoing evaluations or before and after studies generally utilize measured parameters. *Economic benefit* analysis for transportation systems is often the key component of a design evaluation. It usually features a benefit vs. cost analysis [13]. The benefit-to-cost-ratio is often a key indicator for the selection of the design alternative and for the decision to implement the project. Cost benefit analysis is discussed in Chap. 10.

Simple benefit-to-cost ratio comparisons, however, often fail to provide a complete basis for decision making. For example, consider Alternative 1 and Alternative 2 in the hypothetical example of Table 3.4. Although the benefit-to-cost ratio for Alternative 1 is better, if the project objective is to save 600,000 vehicle hours,

Table 3.3 Measures suggested for prospective ITS evaluation

Category	Measure	Units
Economic benefits	Reduction in passenger vehicle delay	Vehicle hours per year
	Reduction in commercial vehicle delay	Vehicle hours per year
	Reduction in commercial vehicle inventory delay	Vehicle hours per year
	Reduction in fuel consumption	Gallons per year
	Reduction in accidents	Accidents per year
Environmental	Carbon monoxide	Pounds per year
	Oxides of nitrogen	Pounds per year
	Volatile organic compounds	Pounds per year
	Carbon dioxide	Pounds per year
User satisfaction	Satisfaction with ITS treatment	KS•NT

KS is a satisfaction rating on a scale of –1.0 to +1.0.
NT is the number of travelers to whom the issue applies.

Table 3.4 Example of marginal analysis

	Alternative 1	Alternative 2	Marginal parameters of alternative 2 relative to alternative 1
Annual vehicle hours saved	250,000	500,000	250,000
Annual benefit	\$5M	\$10M	\$5M
Annualized project cost	\$1M	\$2.5M	\$1.5M
Benefit-to-cost ratio	5:1	4:1	3.33:1
Net benefit	\$4M	\$7.5M	\$3.5M
% of project objective satisfied	41.6	83.3	41.6

Alternative 2 comes much closer to satisfying this objective while maintaining a satisfactory marginal benefit-to-cost ratio (i.e., significantly in excess of 1) for the additional investment. The introduction of the concepts of *marginal values* and *percentage of objective satisfied* provides an important basis for project evaluation.

Travel time variation is an increasingly popular ITS measure. However because of the difficulty in obtaining this quantity in a manner other than direct measurement, it is more suitable for retrospective evaluation rather than prospective evaluation.

While some projects or agencies ascribe monetary values to *environmental benefits*, these benefits may also be a key project objective, as they reduce emissions. This is particularly important in locations that do not conform to the national ambient air quality standard.

The economic and environmental measures do not account for the change in manifest demand that results from some ITS treatments. *Throughput* measures the facility's capability to service demand [14]. Sometimes termed *productivity* [15], it represents the vehicle miles that a facility can accommodate during a peak hour or peak period. Since the effect of the project on the change in this measure is difficult to anticipate in advance of implementation, the measure is best used for retrospective evaluations.

Since the user is the ultimate judge of the perception of ITS effectiveness and value, the system designer must consider the factors that the user deems important. While these factors may be difficult to estimate on a prospective basis, previous studies provide some indication as to how *user satisfaction* will be rated. While Chaps. 5 though 8 discuss user satisfaction for different ITS treatments, some general issues are discussed below.

ITS treatments such as motorist information and motorist service patrols are generally well received by the public because motorists feel that these treatments improve their trip reliability and their sense of control [16]. Rating scales based on surveys may be used to quantify user satisfaction for these types of ITS treatments.

Other treatments such as ramp metering and road pricing or restricted road use (not covered in this book) may provide improved mobility for some users, and, on balance, for the whole system, while providing lower mobility for other users. The transportation system operator must consider how prospective treatments affect different users. Levinson et al. [15] describe an approach to measuring *equity*. The Lorenz Curve (Fig. 3.5) identifies the relationship between the proportion of delay and the proportion of vehicles incurring the delay. AT denotes the area under this curve. Area AD in the figure identifies the users that are relatively disbenefitted by the treatment. The Gini coefficient is computed as

$$G = AD/(AD + AT) \tag{3.3}$$

It quantifies the level of inequality among users. Reference [15] describes a methodology for computing the Gini coefficient.

As an example of the importance of user satisfaction in ITS design and operation, the control strategy for ramp meter rates in Minnesota was revised in order to provide lower delays to entering vehicles, even at the expense of overall increases in delay [15].

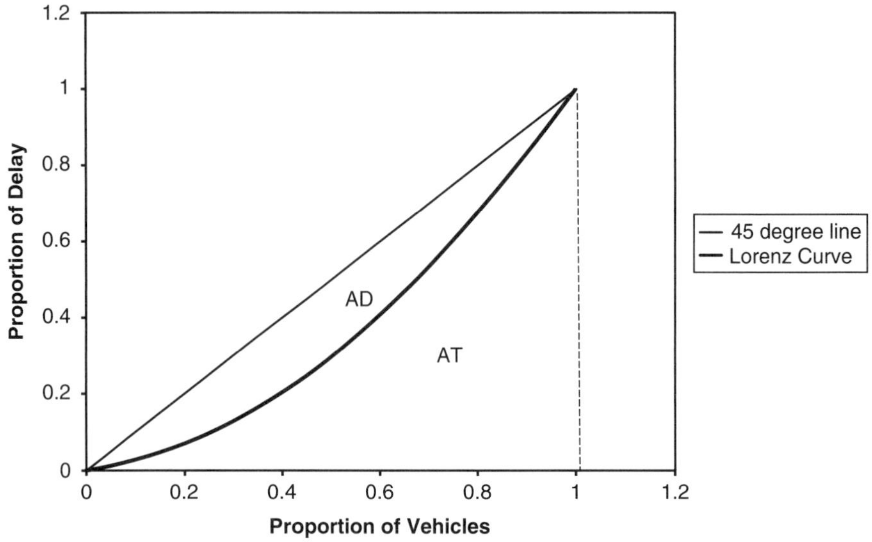

Fig. 3.5 Example of Lorenz Curve for a metered freeway entry ramp

3.2.2 Performance Measures and ITS Planning

Planning agencies such as metropolitan planning organizations (MPOs) are required to select and prioritize projects from a number of alternatives that may include different types of projects and, perhaps, different traveler classes. These projects may affect benefit categories such as those shown in Table 3.3 in different ways. *Multi-attribute utility analysis* is one methodology for comparing such projects [17, 18]. It enables stakeholders with different interests and responsibilities to proactively participate in the planning process. A simple example that compares and depicts the benefits from a candidate set of ITS projects is discussed below.

This analysis uses two types of variables. RI_A denotes the relative importance of attribute A and U_{AS} represents the utility of alternative S for attribute A. RI_A may be developed by stakeholder consensus. The sum of these values for all attributes must equal unity. The attributes in the example are described by the categories shown in Table 3.3.

Table 3.5 lists the parameter values in parentheses under the attribute and utility designations. The second row of the table identifies the attribute value for the example.

The first attribute column in Table 3.5 represents the reduction in traveler cost. Delay, fuel cost, and accidents are major components of this cost. The second attribute column represents emission reduction As described in Chap. 4, for many ITS treatments emission reduction is proportional to motorist delay. The third attribute column is the utility for traveler satisfaction with the treatment.

Table 3.5 Utility values for selected ITS classes

Representative ITS management concepts (S)[a]	Attribute (RI_A)		
RI_A	1. Traveler cost reduced (RI_1=0.6)	2. Emissions reduced (RI_2=0.2)	3. Traveler satisfaction (RI_3=0.2)
1. Improved incident clearance	High reduction (U_{11}=0.9)	Proportional to fuel reduced (U_{21}=0.9)	Little effect observed by motorists (U_{31}=0.2)
2. Restrictive ramp metering	Moderate to high reduction (U_{12}=0.6)	Proportional to fuel reduced (U_{22}=0.6)	May be significant opposition because of ramp delays and local impacts resulting from diverted traffic (U_{32}=0.1)
3. Non-restrictive ramp metering	Lower than Candidate 2 (U_{13}=0.3)	Proportional to fuel reduced (U_{23}=0.3)	Reduced opposition compared to Candidate B (U_{33}=0.3)
4. Motorist information	Moderate (U_{14}=0.4)	Proportional to fuel reduced (U_{24}=0.4)	Generally well accepted by public (U_{34}=0.8)
5. Motorist service patrol	Moderate (U_{15}=0.3)	Proportional to fuel reduced (U_{25}=0.3)	Well accepted by public (U_{35}=0.9)
6. Real time on-board and transit stop information	Little impact (U_{16}=0.1)	Little impact (U_{26}=0.0)	Well accepted by transit passengers (U_{36}=0.5)
7. Transit signal priority	Modest impact (U_{17}=0.3)	Little impact (U_{27}=0.1)	Transit time reliability improvement appreciated by transit users. Little impact on motorists. (U_{37}=0.3)

[a]The candidate alternatives, except for candidates 6 and 7 are discussed in later chapters

Examples of utility values (U_{AS}) are shown in parentheses in the cells that denote the ITS treatments. Values for these measures may be obtained by simulation techniques or inferred from evaluations performed by other ITS projects. Maccubbin et al. [19] provides a compilation of ITS benefits.

The multi-attribute utility values for each alternative S (MU_S) are given by

$$MU_S = \sum_{A=1}^{3} RI_A \cdot U_{AS} \tag{3.4}$$

The multi-attribute utility analysis values for each alternative for the example are given in Table 3.6.

Table 3.6 Multi-attribute utility values for ITS alternatives example

Alternative ITS project classes (S)	Multi-attribute utility value (MU_S)
1. Improved incident clearance	0.76
2. Restrictive ramp metering	0.50
3. Non-restrictive ramp metering	0.30
4. Motorist information	0.48
5. Motorist service patrol	0.42
6. Real time on-board and transit stop information	0.16
7. Transit signal priority	0.26

In most cases the utilities for alternatives selected for this analysis may be added if more than one alternative is chosen for inclusion, in some cases (e.g. candidates 2 and 3) a choice between alternatives is required.

The cost of alternatives may be treated in the following ways in connection with utility analysis:

- As an attribute. This can be done by the addition of a column in Table 3.5.
- As a divisor for the MU_S values in Table 3.6.
- As an abscissa value in a utility vs. cost plot.

In a number of cases, the alternatives may share certain ITS components, therefore the cost for a combination of alternatives may be lower than for each alternative taken separately. Thus it is often desirable to define alternatives that take advantage of this. For example, alternatives 1 and 4 share components such as communications and management centers.

3.3 Alternatives for Functional Design

Design alternatives are required because

- Systems engineering methodologies [20] generally require an alternatives analysis to insure that the design options are considered in the context of project objectives and constraints. A "do nothing" alternative is also usually considered and often constitutes the baseline of benefits for the various design alternatives.
- To obtain federal aid for a project, Title 23, Section 940.11 of the Code of Federal Regulations requires the completion of a systems engineering analysis that includes alternative systems configurations and technology options to meet requirements.
- Many of the states and other agencies responsible for the design and operation of highway projects and ITS require an alternatives analysis in their project design process (see Reference [21] for example).

Two classes of alternatives may be considered:

- Alternative high level ITS project classes

As discussed in Sect. 3.2.2, constraints on regional ITS budgets may require the selection of a subset of projects from the set of alternatives. The first column of Table 3.6 describes the alternatives that were considered for that example.

- Project class design alternatives

Alternatives may include the types of technologies to be included, the intensity of implementation (e.g. how many point detectors to include in a roadway section, and functional equipment placements (e.g. the best location for a changeable message sign).

3.3.1 Design Constraints

Design constraints limit the selection of components and operations that are suitable for the project. Reference [20] provides the following discussion of constraints:

"The fulfillment of goals and the approaches to satisfy specific functional requirements are often constrained by resource, institutional, and legacy issues. In some cases the necessity to resolve problems may justify the relaxation of constraints. In the absence of this situation, the use of constraint analysis has the potential to simplify the selection of design alternatives by eliminating alternatives lying outside constraint boundaries. Some of the more common constraints are listed below.

Resource Constraints

- Capital funding
- Funding for operations
- Funding for maintenance
- Staffing levels and capabilities

Reference [21] identifies constraints as part of the project development process. Federal regulations require that both long range and short range plans be financially constrained to reflect revenues reasonably expected to be available over the time period they cover [22].

Institutional Constraints

- Funding through long term planning processes
- Requirements to use agency specific standard specifications
- Requirements to use National ITS Architecture standards and protocols
- Requirements to provide interoperability with other ITS in the same jurisdiction or other jurisdictions
- General design constraints

- Preservation of existing utilities
- Right-of-way constraints
- Economic, social, environmental, and community considerations

Legacy Constraints

- Requirements to use existing equipment to the extent possible
- Requirements for new equipment to be compatible with existing equipment

Early identification of constraints will result in either or both of the following benefits:

- The potential benefits of the project or design approach indicate that a serious attempt should be made to relieve the constraint.
- The project must be subject to the constraints. These constraints may eliminate some alternatives from further consideration."

3.3.2 Relationship of ITS Management Concepts to Objectives

Chapter 2 described general objectives that may be considered as candidates for a project. Table 3.7 shows how these objectives are related to a number of ITS freeway management concepts. This table may assist in developing candidate concepts and alternatives for the project.

While Table 3.7 shows the types of management concepts available to address objectives, another key functional design issue is the *intensity of deployment* for ITS treatment. Delay increases with volume-to-capacity ratio (as shown in Fig. 4.4) and the capability for a particular ITS treatment to reduce this delay also increases with volume-to-capacity ratio (q/C). Because higher q/C situations provide greater benefits for a given cost, deployments for those cases will provide higher benefit-to cost ratios. This is illustrated conceptually in Fig. 3.6, which shows that, as the volume-to-capacity ratio increases, a given investment not only provides greater benefits but also a level of benefits that cannot be obtained at any investment level for a lower volume-to-capacity ratio.

Reference [21] describes an approach to categorizing traffic levels so that ITS deployment intensities may be selected to implement the concepts of Fig. 3.6. These levels are defined as follows:

- Level 3 – A continuous section of roadway in one direction that includes peak hour level of service D or worse traffic for at least one half of the section.
- Level 2 – A continuous section of roadway in one direction that includes peak hour level of service C or worse traffic for at least one half of the section. Conditions worse than level of service C may be present at scattered spot locations

Table 3.7 Relationship of ITS freeway management concepts to general objectives (redrawn) [21]

Management concept / Objective	Recurrent congestion detection techniques	Non-recurrent congestion detection and incident tracking	Traveler information	Ramp metering	Road weather information system	Motorist assistance	Data collection and management for planning and performance evaluation	Monitoring of ITS equipment and inventory management, security
1. Reduce congestion/ improve travel time								
a. Recurrent congestion – significant section	✓		✓	✓			✓	
b. Recurrent congestion – spot	✓		✓	✓			✓	
c. Non-recurrent congestion (significant section)		✓	✓			✓		
d. Non recurrent congestion (spot)		✓	✓			✓		
2. Reduce accident rate								
a. Over significant section	✓	✓	✓	✓	✓	✓	✓	
b. Spot	✓	✓	✓	✓	✓			
3. Reduce emissions and fuel consumption	✓	✓	✓	✓		✓	✓	
4. Serve as a corridor link in a wider area highway system	✓	✓	✓		✓		✓	
5. Serve as a diversion route in local corridor	✓	✓	✓		✓		✓	

(continued)

Table 3.7 (continued)

Management concept / Objective	Recurrent congestion detection techniques	Non-recurrent congestion detection and incident tracking	Traveler information	Ramp metering	Road weather information system	Motorist assistance	Data collection and management for planning and performance evaluation	Monitoring of ITS equipment and inventory management, security
6. Special traffic management functions								
a. Major roadway reconstruction	✓	✓	✓	✓		✓	✓	
b. Minor roadway construction	✓	✓	✓			✓	✓	
c. High occupancy vehicles		✓	✓	✓	✓	✓	✓	
f. Traffic information on roadway construction	✓	✓	✓				✓	
g. Information on weather conditions, parking, special events, road weather		✓	✓		✓		✓	
j. Traffic management for future major roadway reconstruction	✓	✓	✓				✓	
k. Motorist information about detour routes			✓					
7. Interoperability of ITS								
a. Operator efficiency	✓	✓	✓		✓			
b. Stakeholder involvement	✓	✓	✓	✓	✓	✓	✓	✓

(continued)

Table 3.7 (continued)

Management concept / Objective	Recurrent congestion detection techniques	Non-recurrent congestion detection and incident tracking	Traveler information	Ramp metering	Road weather information system	Motorist assistance	Data collection and management for planning and performance evaluation	Monitoring of ITS equipment and inventory management, security
8. Improvement of NYSDOT Operations								
a. Planning and/ or evaluation data collection	✓			✓	✓		✓	
b. ITS equipment monitoring							✓	✓
c. Efficiency of operations					✓			✓
d. Reduction of operating and/or maintenance cost					✓		✓	✓
9. Provide assistance to disabled motorists		✓				✓		
10. Provide travel information related to tourism			✓		✓			
11. Improve security								
a. Transportation system security								✓
b. Emergency operation		✓	✓		✓			✓
c. Information system security								✓
12. Improve commercial vehicle operations	✓	✓	✓		✓			

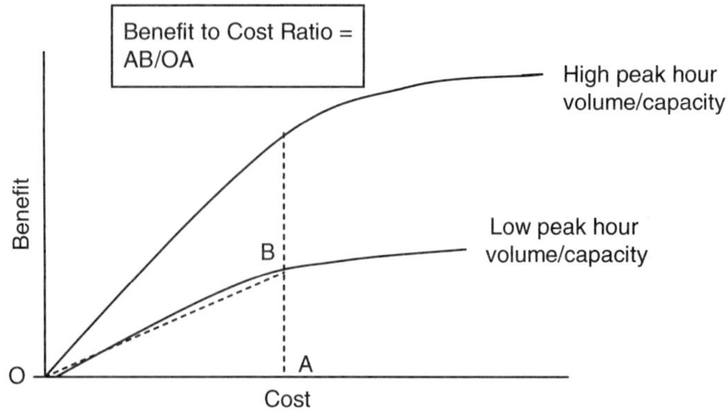

Fig. 3.6 Benefits and costs for ITS deployments

Table 3.8 Representative implementation characteristics for freeway ITS

Capability	Level 1 – minimal	Level 2 – moderate	Level 3 – intensive
TMC site and staffing	Minimal site cost. May be part time or partial permanent staff	Moderate site cost. May be part time or full time staffing	Full time staffing
Computer system for central management of key ITS functions	May be a computer to provide a low level of management capability	Usually	Yes
CCTV coverage	Minimal	Significant	Full
Roadway mainline detector complement (point or probe detectors)	Typically none	Not continuous	Usually continuous
Changeable message signs	Appropriate locations, possibly at major diversion points	Diversion locations and other key locations	Diversion locations, periodic intervals, possibly at key entry locations
Service patrols	Sometimes	Often	Yes
Ramp metering	None	Rarely	Frequently

or for small sections. In this case, it may be appropriate to increase the concentration of field equipment at these locations.

- Level 1 – Traffic conditions better than Level 2.

Table 3.8 [21] indicates the general deployment intensity of ITS devices or operations commonly implemented by many ITS that generally conform to the benefit-cost concepts of Fig. 3.6.

Chapters 4 and 5 provide more detailed application factors for equipment deployment as a function of traffic level.

3.4 Evaluation Concepts

The following sections discuss some of the tools and techniques that may assist in establishing the values of benefits for benefit vs. cost analysis. Benefit vs. cost analysis is described in more detail in Chap. 10.

3.4.1 Benefit and Cost Modeling

A number of models exist that provide an estimate of benefits for ITS design alternatives. All of the models contain default factors for benefits computation that may be altered by the user. The accuracy of the models is significantly influenced by the extent to which the factors represent the actual scenario. Therefore the credibility of the results increases when more conservative factors are chosen.

Information sources for benefits factors include:

- Prior experience in the region
- Research reports
- The DOT ITS benefits database [19]

While some of the models provide cost information, it is recommended that cost default factors provided by the model not be used because the information is usually out of date and may not sufficiently represent actual site conditions.

Four models are described below. Details on how these models may be acquired are included in Appendix A.

3.4.1.1 Design ITS

This model provides evaluation of freeway ITS and is implemented by a detailed Excel-based workbook. The model provides improved benefits estimations compared to the earlier EMFITS model (Sect. 3.4.1.1) and facilitates the development of design alternatives and tradeoffs for a number of ITS devices. It estimates benefits for the following ITS functions:

- Improvement in recurrent congestion resulting from motorist information.
- Improvement in non-recurrent congestion resulting from motorist information.
- Improvement in non-recurrent congestion resulting from reduction of incident clearance time.
- Improvements resulting from ramp metering.
- Improvements resulting from motorist service patrols.

3.4.1.2 Evaluation Model for Freeway ITS Scoping (EMFITS)

This public domain model was developed for the New York State Department of Transportation. It is an earlier version of the Design ITS model that is less detailed in some respects and that does not have the detailed equipment placement guidance provided by the Design ITS model.

3.4.1.3 ITS Deployment Analysis System (IDAS)

This model was developed for the Federal Highway Administration. It provides benefits and cost evaluation capability for a number of ITS initiatives other than freeway ITS improvements. The model is best used in conjunction with a regional traffic planning model and accepts traffic data from several such models.

3.4.1.4 Screening Analysis for ITS (SCRITS)

This public domain model is an early Excel workbook based model developed for the Federal Highway Administration. It provides benefits and cost evaluation capability for a number of ITS initiatives other than freeway ITS improvements as well as for ITS treatments.

3.4.2 Benefit Paradigms for ITS Evaluation Models

Evaluation models require a process or paradigm for developing benefit measures for the effect of the physical ITS devices and the management strategies that they implement. The Design ITS model employs the paradigm shown in Fig. 3.7.

The thread structure in Fig. 3.7 provides a framework for computing benefits. Each type of line in the figure represents a separate thread group. For example, the heavy solid line identifies the threads originating from incident management information to motorists (A2). This information results in diversion (B1) which, when properly implemented, reduces non-recurrent congestion (C2). This reduction results in time savings (M1) and fuel and emissions reduction (M3). Thus incident information to motorists generates thread groups A2B1C2M1 and A2B1C2M3. The evaluation model computes benefits for all of the threads identified by this process.

The lowest tier (Tier A) represents the physical ITS devices and management strategies. Tier B represents the physical mechanism for achieving benefits. Tier C represents the effects of diversion as a result of motorist information. Tier M exhibits the classes of benefit measures provided by Design ITS. The thread groups originating from Tier A are:

- Recurrent congestion information to motorist (Thread Group A1)
- Incident information to motorist (Non-recurrent congestion) (Thread Group A2)
- ITS facilitation of incident clearance (Thread Group A3)
- Ramp metering (Thread Group A4)
- Service patrols (Thread Group A5)

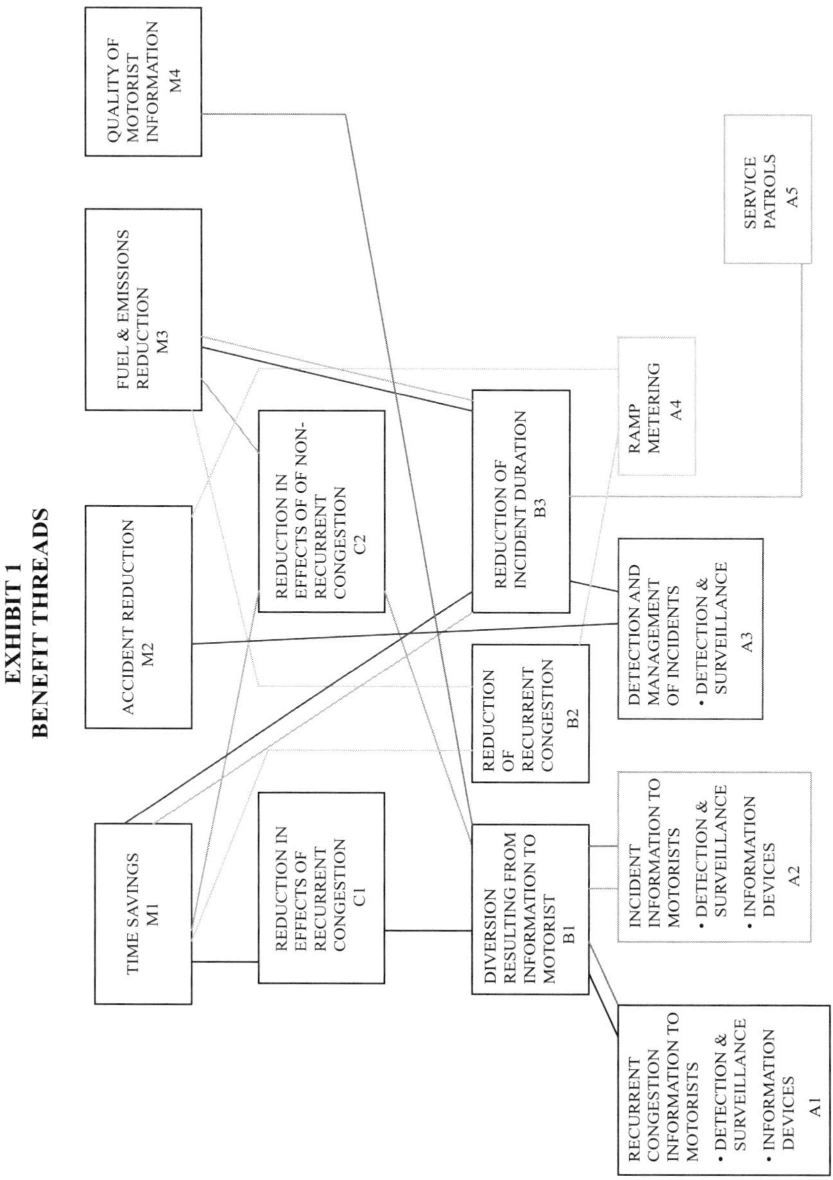

Fig. 3.7 Example of evaluation model paradigm [23]

Design ITS annual benefit measures (Tier M) include the following:

- Vehicle hours of delay saved*
- Accidents reduced*
- Gallons of fuel reduced*
- Grams of hydrocarbons reduced
- Grams of carbon monoxide reduced
- Grams of oxides of nitrogen reduced
- Ton hours of goods delay reduced*
- Relative quality of motorist information

In addition to the quantities themselves, dollar values for the measures indicated by an asterisk (*) are provided by Design ITS.

The model does not assume that the benefits due to multiple sources of motorist information (e.g. the combined use of CMS and HAR) are entirely cumulative. It models the interaction of multiple media. Similarly, Design ITS accounts for the interaction of redundant or overlapping detection and surveillance technologies [23]. Relationships for modeling these interactions are described in later chapters.

References

1. May AD (1990) Traffic flow fundamentals. Prentice-Hall, Englewood Cliffs
2. Wardrop JG Some theoretical aspects of road traffic research. Road Paper 36, Proceedings Institution of Civil Engineers, Pt 2, 1:325–378
3. Drake JS, Shofer JL, May AD Jr. (1967) A statistical analysis of speed-density hypotheses. Highway Research Record 154, 53–87
4. Highway Capacity Manual. (2000) Transportation Research Board, National Research Council, Washington, DC
5. Lomax T, et al (1997) Quantifying congestion. NCHRP Report 398, National Academy Press, Washington DC
6. Peeta S et al (2000) Content of variable message signs and on-line driver behavior. Transportation Research Record No. 1725:102–108
7. Traffic Assignment Manual (1964) Bureau of Public Roads. U.S. Department of Commerce, U.S. Government Printing Office
8. Ullman GL, Dudek CL, Balke KN (1994) Effect of freeway corridor attributes upon motorist diversion responses to travel-time information. Transportation Research Record No. 1464: 19–27
9. Dudek CL (2004) Changeable message sign operation and messaging handbook. Report No. FHWA-OP-03-070, Federal Highway Administration, Washington, DC
10. Wardrop JG (1952) Some theoretical aspects of road traffic research. Proceedings, Institute of Civil Engineers, Part II 1:325–378
11. Neudorff LG et al (2003) Freeway management and operations handbook. Administration Report FHWA-OP-04-003, Federal Highway Administration, Washington, DC
12. Shaw T (2003) Performance measures of operational effectiveness for highway segments and systems – a synthesis of highway practice. NCHRP Synthesis 311, Transportation Research Board, Washington DC
13. Lee DB Jr (2004) Making the case for ITS investment. In: Gillen D, Levinson D (eds) Assessing the benefits and costs of ITS. Kluwer Academic Publishers, Norwell

14. Gordon RL et al (1995) Traffic control systems handbook. Report FHWA-SA-95-032, Federal Highway Administration, Washington, DC
15. Levinson D et al (2004) Measuring the equity and efficiency of ramp meters. University of Minnesota, Minneapolis, MN
16. Brand D (2004) Benefit measures, values, and future impacts of ITS. In: Gillen D, Levinson D (eds) Assessing the benefits and costs of ITS. Kluwer, Norwell, MA
17. Wang Z, Walton CM (2008) A Multi-attribute utility theory approach for ITS planning. 87th Annual Meeting, Transportation Research Board, Washington, DC
18. Levine J, Underwood SE (1996) A multiattribute analysis of goals for intelligent transportation systems planning. Transportation Research-C 4(2):97–111
19. Maccubin RP et al (2003) Intelligent transportation systems benefits and costs: 2003 update. Report FHWA-OP-03-075, Mitretek Systems, Inc., Washington, DC
20. Gordon RL (2003) Systems engineering processes for developing traffic signal systems. NCHRP Synthesis 307, Transportation Research Board, Washington, DC
21. Project development manual, Appendix 6. New York State Department of Transportation. https://www.nysdot.gov/portal/page/portal/divisions/engineering/design/dqab/pdm. Accessed 1 March 2009
22. Smith SA (1998) Integrating intelligent transportation systems within the planning process: an interim handbook. Report FHWA-SA-98-048, Federal Highway Administration, Washington, DC
23. Gordon RL (2007) Design ITS user's manual

Chapter 4
Non-Recurrent Congestion: Improvement of Time to Clear Incidents

Abstract The greatest benefit that freeway ITS can provide is to assist in the reduction of the time that it takes emergency responders to clear incidents and restore normal traffic operations. This chapter describes the ways that ITS can facilitate these operations and the benefits that result. Guidance for the field location of ITS components is provided. The chapter covers the following:

- Stages of an incident
- Effects of an incident on roadway capacity and models of delay resulting from an incident
- Relationship of incident clearance time to delay
- Adaptation of delay models to local traffic data
- Design functions and technologies to assist in incident management
- CCTV coverage requirements and camera placement guidelines
- Traffic detector technologies and placement guidelines
- Improvement of traffic management center support of incident management
- Evaluation of incident management effectiveness

4.1 Description of an Incident for ITS Design Purposes

A traffic incident is an unplanned, non-recurring activity on or near the roadway that causes a reduction of roadway capacity or an abnormal increase in demand. Such events include traffic accidents, disabled vehicles, and spilled cargo. Highway maintenance and reconstruction projects are sometimes considered as incidents, but our definition excludes these activities. Emergencies such as natural disasters and terrorist attacks are also unplanned and they also can cause a reduction of capacity or an abnormal increase in demand. Their impacts and management requirements, however, extend well beyond the roadway [1].

The following major stages of incident management include [1]:

- Detection – Determining that a traffic incident has occurred.
- Verification – Determining the precise location and nature of an incident, as well as the display, recording, and communication of this information to the appropriate agencies.

R. Gordon, *Intelligent Freeway Transportation Systems: Functional Design*,
DOI 10.1007/978-1-4419-0733-2_4,

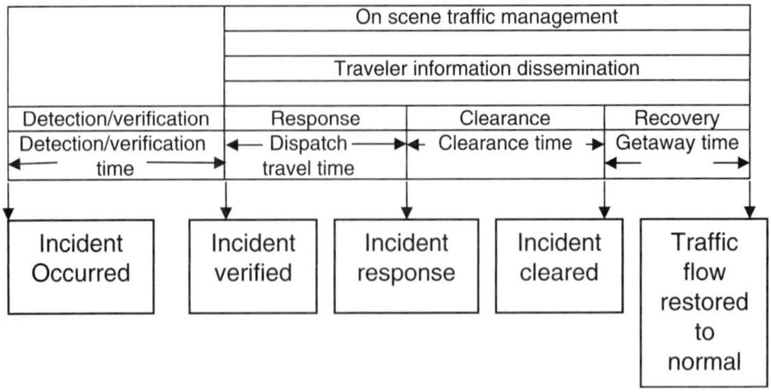

Fig. 4.1 Stages of incident management [2] (redrawn)

- Response – Activation, coordination, and management of the appropriate personnel, equipment, communication links and motorist information media as soon as it is reasonably certain that a traffic incident has occurred.
- Clearance – Removal of vehicles, wreckage, debris, spilled material, and other items from the roadway and the immediate area to restore roadway capacity.
- Recovery – Restoring traffic flow at the site of the traffic incident; preventing more traffic from flowing into the area and getting trapped in the upstream queue; and preventing congestion from spilling across the roadway network.

These stages are shown in Fig. 4.1. The ITS issues for incident management are discussed in Sect. 4.5.

4.1.1 Effect of Incidents on Capacity

The effect of capacity reduction on a freeway is far greater than the physical reduction of roadway width. This effect is shown by Lindley [3] in Table 4.1.

4.1.2 Secondary Accidents

Secondary accidents are accidents that result from an existing primary incident. Many times, these accidents occur at the end of queues that result from the primary incident. Raub [4] estimates that more than 15% of the crashes reported by the police may be secondary accidents. Reducing the duration of queues resulting from incidents not only reduces delay to motorists due to the incident, but also results in a reduced rate of secondary accidents, and a consequent reduction in the total accident rate.

Table 4.1 Fraction of freeway section capacity available under incident conditions

Number of freeway lanes in each direction	Shoulder disablement	Shoulder accident	Lanes blocked		
			One	Two	Three
2	0.95	0.81	0.35	0	N/A
3	0.99	0.83	0.49	0.17	0
4	0.99	0.85	0.58	0.25	0.13
5	0.99	0.87	0.65	0.40	0.20
6	0.99	0.89	0.71	0.50	0.25
7	0.99	0.91	0.75	0.57	0.36
8	0.99	0.93	0.78	0.63	0.41

Source: From Transportation Research Record 1132, Transportation Research Board, National Research Council, Washington, DC, 1987, Table 1, P.6. Reproduced with permission of the Transportation Research Board. These data appear in later sections in conjunction with demand volumes to provide estimates of delay resulting from incidents

4.2 Models of the Effects of Freeway Incidents

Deterministic queuing models are often used to analyze the delay and timeline associated with freeway incidents that restrict the capacity to below the demand volume. An example of a simple model is shown in Fig. 4.2. While models of this type do not capture all of the details of the dynamic traffic flow characteristics, they provide approximations that are useful for design and evaluation purposes.[1]

The time for the queue to dissipate after the incident is cleared is given by

$$T_D = (q_2 - q_3) \bullet T / (q_1 - q_2) \tag{4.1}$$

where

q_1 = Volume at incident clearance (roadway capacity)
q_2 = Volume entering incident location (demand volume)
q_3 = Volume when incident is present (restricted capacity resulting from incident)
T = Time from start of incident to incident clearance

The line q_2 in the figure represents the number of vehicles that enter the incident location while $q_1$and q_3 depict the number of vehicles that are released from the incident location. The vertical distance between q_2 and either q_3 or q_1 represents the number of vehicles in the queue. QC is the queue length at the time the incident is cleared. D_I, the delay until incident clearance, is calculated as

$$D_I = (q_{2-}q_3) \bullet T^2 / 2 \tag{4.2}$$

[1]This model and the related model discussed in Sect. 4.3 do not incorporate the delay resulting from the additional vehicles delayed by the upstream propagation of the tail of the queue.

Fig. 4.2 Delay and timeline model for incidents

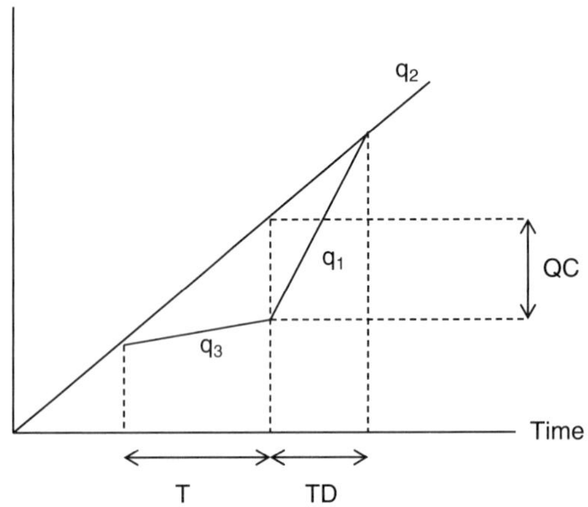

The delay from incident clearance to queue dissipation is given by

$$D_Q = (q_2 - q_3)^2 \bullet T^2 / (2 \bullet (q_1 - q_2)) \tag{4.3}$$

Total delay is:

$D_T = D_I + D_Q$

$$D_T = (q_2 - q_3) \bullet T^2 / 2 + (q_2 - q_3)^2 \bullet T^2 / [2 \bullet (q_1 - q_2)] \tag{4.4}$$

A more complex model that includes a period of total closure is provided by Morales [5].

Figure 4.3 shows an example of incident delay as a function of the time to clear an incident blocking one lane of a roadway with three lanes in one direction for a typical peak hour condition without recurrent congestion. Figure 4.4 shows the strong sensitivity of incident delay to volume-to-capacity ratio for the freeway upstream of that incident.

There is considerable variation in the accident clearance time data provided by different agencies. Ozbay and Kachroo [2] provide data obtained in Northern Virginia for different types of incidents. For example, an analysis of lane blocking incident data on freeways in Long Island, N.Y, prior to the installation of ITS showed an average time to clear as 49.6 min, while data for similar conditions in the Atlanta area prior to ITS installation showed 64 min [6].

Sections 4.1 and 4.2 show that the delay corresponding to an incident varies significantly with the number of lanes that the facility has, the lane blocking effects of the incident, the time for the incident to be cleared, and the demand volume in

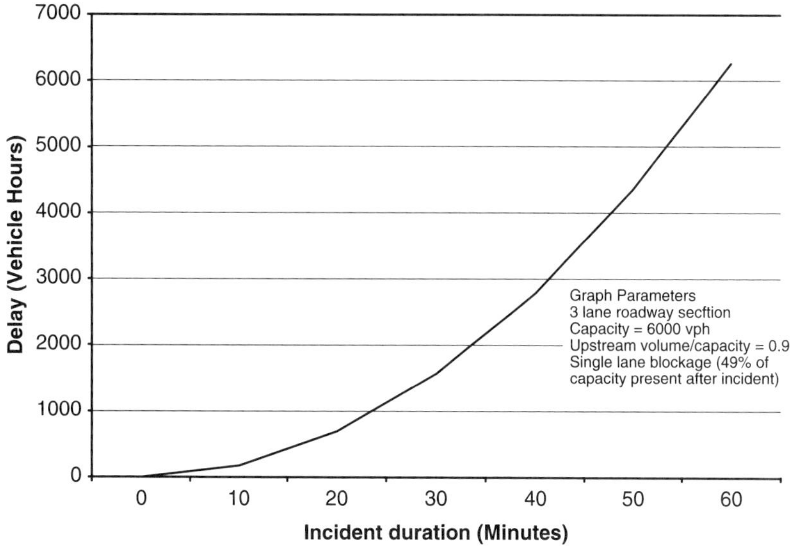

Fig. 4.3 Vehicle hours of delay vs. time to clear incident

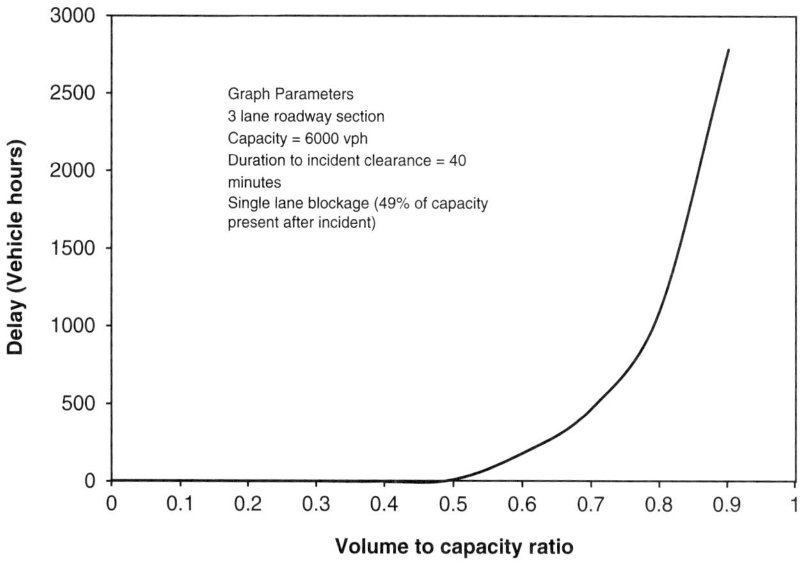

Fig. 4.4 Delay vs. volume to capacity ratio upstream of incident

the absence of an incident. The practical evaluation of ITS improvements requires a relatively simple model that embeds these variations. Sections 4.3 and 4.4 describe an approach to the development of such a model.

4.2.1 Frequency and Severity of Incidents

The evaluation model in Sect. 4.4 requires a value for the blockage time and severity (capacity reduction) of a representative incident. It also requires a value for the frequency of the incident (number of incidents per million vehicle miles of travel). Figure 4.5 shows a model that provides some of the required information.

The model does not contain information on incident frequency. Furthermore, the model shows wide variations in the incident duration. Since a good deal of this variation is likely due to differences in reporting styles and the definition

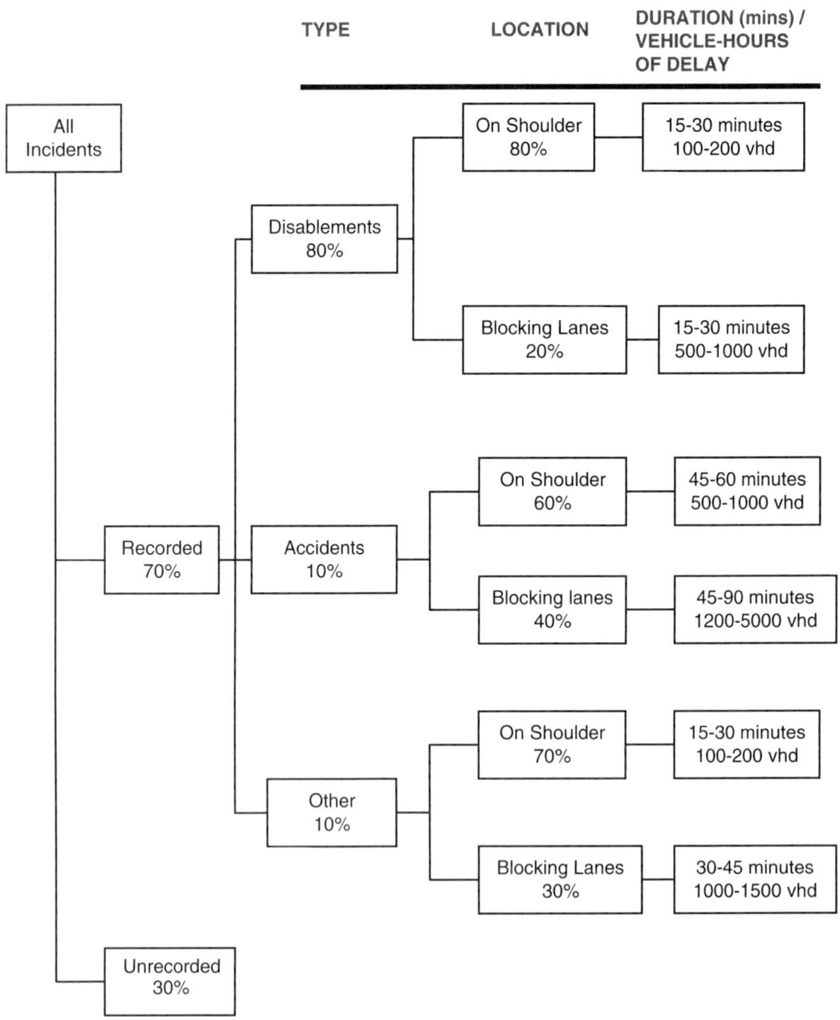

Fig. 4.5 Composite profile of reported incidents by type [7]

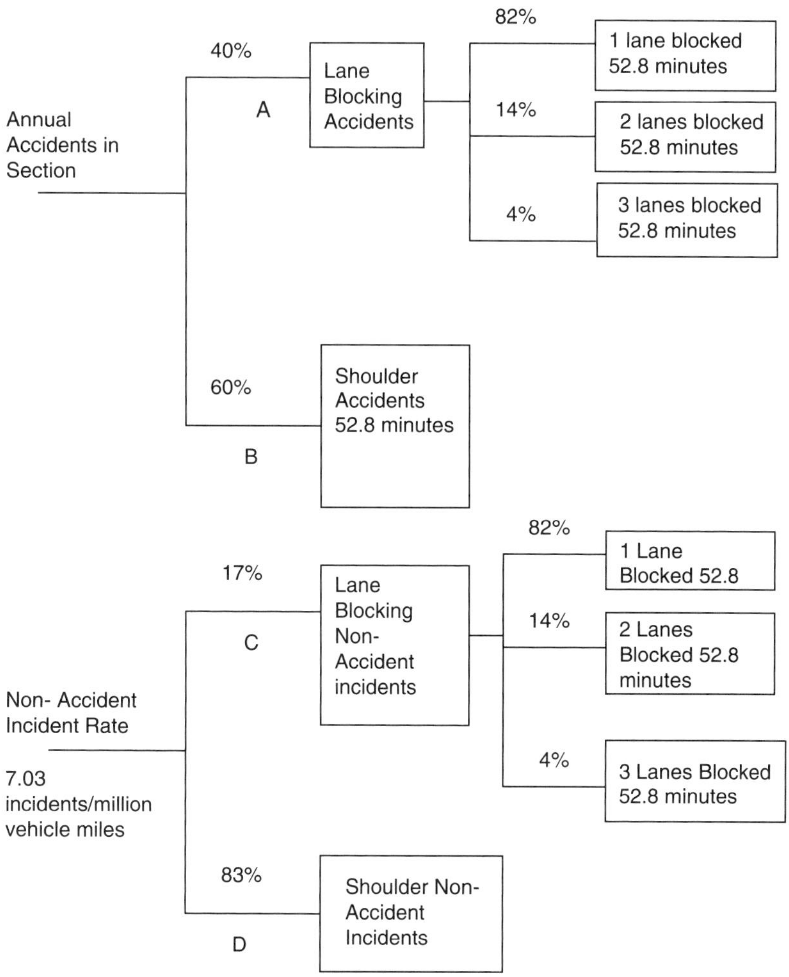

Fig. 4.6 Incident model structure

of an incident, a more appropriate approach for practitioners is to provide a model and procedure that may be calibrated on a local or area-wide basis. The model structure in Fig. 4.6 was developed by the New York State Department of Transportation, and provides a framework for agencies responsible for freeway management operations to calibrate the model using locally obtained data [8]. The model treats accidents and nonaccident incidents separately, thus, enabling commonly reported accident rate data to be used directly. The model considers a section as the roadway portion between the centers of two interchanges. The data shown in the figure are based on observations in several upstate New York metropolitan areas. The limited data sample was not sufficient to differentiate among the lane blockages.

4.2.2 Data Collection for Development of Incident Model

Experience has shown that in many cases, historic data from the incident logs commonly provided by traffic management systems are often not provided with sufficient fidelity to enable nonaccident incident rates and durations to be obtained. For example, the development of Fig. 4.6 required the documentation of data in more detail and in a different format than that provided by normal operation of the traffic management system. It is, therefore, recommended that a separate data set be collected for this purpose by traffic management center operators who have been briefed on the model and its data collection requirements. This data set should be based on CCTV observations.

Data collection will be required for a number of sections to provide a sufficient data sample for analysis. The parameters required for analysis include

- Roadway Section ID
- Incident ID
- Roadway direction
- Date
- Time incident detected
- Time incident cleared
- Accident or nonaccident incident
- Shoulder or moving-lane incident
- Number of lanes closed
- Duration of lane closure

4.3 Relationship of Reduction in Delay to Reduction in Incident Clearance Time

The area abcd in Fig. 4.7 shows the effect on delay of a reduction in the time to clear the incident.

Equation 4.4 shows that the total delay is proportional to the square of the time interval from the start of the incident until incident clearance. Thus, (4.4) may be rewritten as

$$D_T = K \bullet T^2 \tag{4.5}$$

The effect on delay of small changes in the time to clear the incident is given by the derivative of D_T with respect to T as

$$dD_T / dT = 2 \bullet K \bullet T \tag{4.6}$$

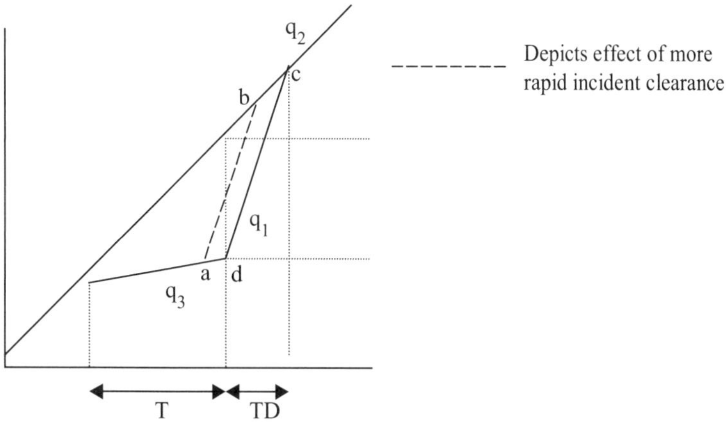

Fig. 4.7 Delay reduction resulting from reduction in incident clearance time

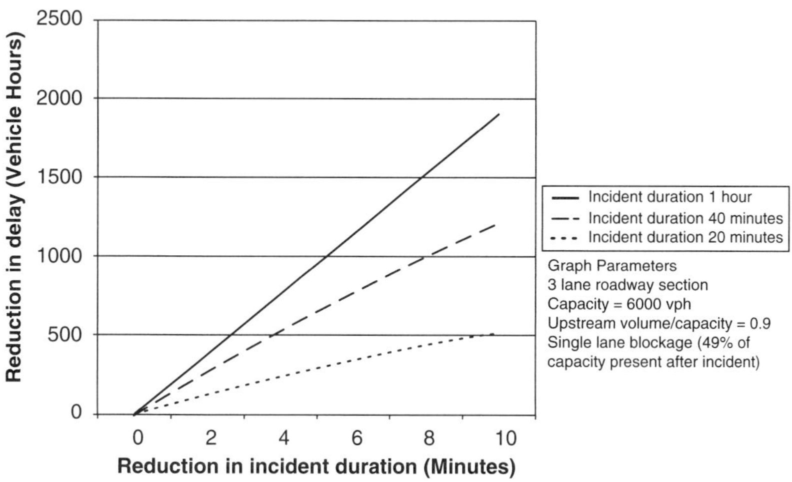

Fig. 4.8 Example of peak period reduction in delay as a result of reduction in time to clear incident

The ratio of change in delay to incident clearance time is

$$(1/T)\bullet(dD_T/dT) = 2\bullet K \tag{4.7}$$

From this equation, it is seen that a small percentage change in the reduction in the time to clear the incident results in twice that percentage of delay reduced.

An example of the reduction in delay due to reduction in incident clearance time is shown in Fig. 4.8. It was computed using (4.4) by taking the difference of the delay prior to and after the reduction in incident clearance time. A similar graph for off-peak period incidents is shown in Fig. 4.9.

Fig. 4.9 Example of off-peak period reduction in delay as a result of reduction in time to clear incident

4.4 Interaction of Capacity Restrictions and Traffic Conditions

Because an incident may occur at any time, the impact on delay will depend upon the volume at that time and the residual capacity as shown in Table 4.1.

Incident models of the types shown in Figs. 4.5 and 4.6 provide the basis for assigning a probability of occurrence to each blockage time. A general plan for developing a model for average incident delay is shown in Fig. 4.10. The plan depicts an approach that groups the likelihood of incidents into volume-to-capacity ratio based groups or "cohorts" that represent the effect of various types of incidents. Delay is computed for each cohort, and then the cohorts are assembled into the final delay estimate. Sections 4.4.1 and 4.4.2 discuss the details of the approach.

4.4.1 Cohort Model

In order to estimate the percentage of traffic at each volume-to-capacity level, the representative hourly traffic volumes may be classified into cohorts designed to capture the effect of the various lane blockage scenarios [8]. For example, for a freeway with three lanes in each direction, the cohorts for a single lane blockage may be classified as shown in Table 4.2. The cohort range definitions (0.49 and 0.17) correspond to the blocked lane capacities shown in Table 4.1.

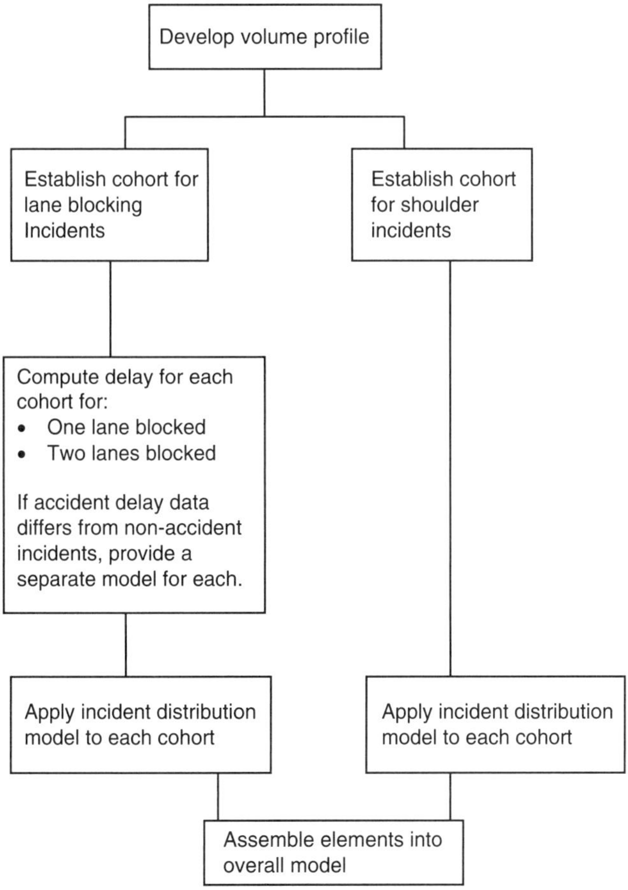

Fig. 4.10 Flow diagram to develop incident delay

Table 4.2 Cohorts for freeway with three lanes in each direction

Cohort number	Volume to capacity ratio range in cohort	Representative volume to capacity ratio used in analysis
1	$0.7 \geq q/C$	0.8
2	$0.7 \geq q/C \geq 0.49$	0.6
3	$0.49 \geq q/C \geq 0.17$	0.33
4	$0.17 \geq q/C \geq 0$	0.1

The last column of the table provides a volume-to-capacity ratio characteristic of the cohort and used in the computation of delay. Figure 4.11 shows the relationship of these cohorts to a typical weekday freeway volume characteristic. To develop the model, the hourly volumes are assigned to the proper cohort and summed. The data are then used to identify the fraction of the annual average daily traffic (AADT) that is present when different lane blockages occur.

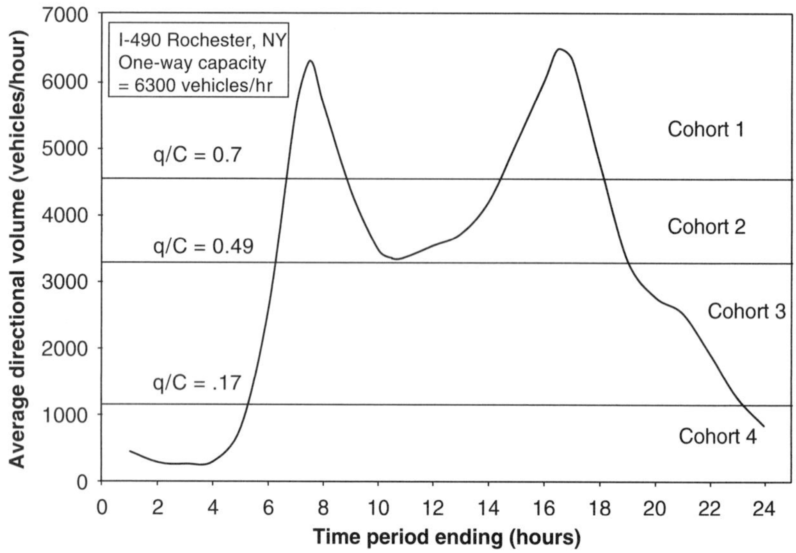

Fig. 4.11 Relationship of cohorts to volume profile

Figure 4.12 shows an example of the worksheet used to develop the fraction of AADT in each for a three-lane roadway.

Hourly volume for each roadway direction is entered into the worksheet along with the capacity for the roadway segment. The volume-to-capacity ratio is computed by the worksheet. The worksheet sorts the volumes into cohorts using the cohort definitions in Table 4.2. The sorted volumes for each hour are then summed over the day to determine the daily volume for each cohort.

Table 4.1 which depicts the percentage of freeway capacity under incident conditions on the shoulder shows that for three lane freeways, a q/C of 0.99 must be present for a disablement type of incident to have an effect. Since this condition is not present for the volume profile of Fig. 4.11, that type of incident was not considered in the development of Fig. 4.12. The table does, however, show that shoulder accidents may develop congestion under some conditions in Fig. 4.11. Thus, the analysis shown in Fig. 4.12 also identifies the fraction of the AADT for which shoulder accidents will influence delay. The volumes in the example represent a situation with no recurrent congestion.

The enclosed CD contains the worksheet file Cohort Factors 3 Lanes file for Fig. 4.12.

4.4.2 Time Saved per Incident

The cohort fraction data, such as in Fig. 4.12, may be used in conjunction with the incident delay model in (4.4) to compute the delay for lane blocking incidents for

COMPUTATION OF COHORT FRACTIONS (THREE LANE ROADWAY)

Data entry required

Roadway Segment					I-490 from I-390 to Mt. Read Blvd.		Cohort Volumes for Lane Blocking Incidents								Cohort Volumes for Shoulder Accidents	
Beginning Hour	Direction			Est Cap	eastbound	westbound										
	Eastbound	Westbound	Total	1 way	v/c	v/c	eb cohort 1	eb cohort 2	eb cohort 3	eb cohort 4	wb cohort 1	wb cohort 2	wb cohort 3	wb cohort 4	eb v/c>=.83	wb v/c>=.83
12:00 AM	284	671	955	6300	0.045	0.107	0	0	0	284	0	0	0	671	0	0
1:00 AM	164	519	683	6300	0.026	0.082	0	0	0	164	0	0	0	519	0	0
2:00 AM	164	490	654	6300	0.026	0.078	0	0	0	164	0	0	0	490	0	0
3:00 AM	222	468	690	6300	0.035	0.074	0	0	0	222	0	0	0	468	0	0
4:00 AM	441	519	960	6300	0.070	0.082	0	0	0	441	0	0	0	519	0	0
5:00 AM	1550	842	2392	6300	0.246	0.134	0	0	1550	0	0	0	0	842	0	0
6:00 AM	3831	2088	5919	6300	0.608	0.331	0	3831	0	0	0	0	2088	0	0	0
7:00 AM	5830	2769	8599	6300	0.925	0.440	5830	0	0	0	0	0	2769	0	5830	0
8:00 AM	4714	2300	7014	6300	0.748	0.365	4714	0	0	0	0	0	2300	0	0	0
9:00 AM	2751	1869	4620	6300	0.437	0.297	0	0	2751	0	0	0	1869	0	0	0
10:00 AM	2182	1832	4014	6300	0.346	0.291	0	0	2182	0	0	0	1832	0	0	0
11:00 AM	2142	2031	4173	6300	0.340	0.322	0	0	2142	0	0	0	2031	0	0	0
12:00 PM	2143	2180	4323	6300	0.340	0.346	0	0	2143	0	0	0	2180	0	0	0
1:00 PM	2201	2170	4371	6300	0.349	0.344	0	0	2201	0	0	0	2170	0	0	0
2:00 PM	2591	3064	5655	6300	0.411	0.486	0	0	2591	0	0	0	3064	0	0	0
3:00 PM	2869	4472	7341	6300	0.455	0.710	0	0	2869	0	4472	0	0	0	0	0
4:00 PM	2956	5282	8238	6300	0.469	0.838	0	0	2956	0	5282	0	0	0	0	5282
5:00 PM	2988	5148	8136	6300	0.474	0.817	0	0	2988	0	5148	0	0	0	0	0
6:00 PM	2151	2896	5047	6300	0.341	0.460	0	0	2151	0	0	0	2896	0	0	0
7:00 PM	1514	1905	3419	6300	0.240	0.302	0	0	1514	0	0	0	1905	0	0	0
8:00 PM	1261	1789	3050	6300	0.200	0.284	0	0	1261	0	0	0	1789	0	0	0
9:00 PM	1222	1659	2881	6300	0.194	0.263	0	0	1222	0	0	0	1659	0	0	0
10:00 PM	870	1207	2077	6300	0.138	0.192	0	0	0	870	0	0	1207	0	0	0
11:00 PM	505	1067	1572	6300	0.080	0.169	0	0	0	505	0	0	0	1067	0	0
Total =	47546	49237	96783			Total	10544	3831	30521	2650	14902	0	29759	4576	5830	5282

Frac shoulder accidents with V/C>= .83 0.115

cohort	fraction
1	0.263
2	0.040
3	0.623
4	0.075

Fig. 4.12 Worksheet for computation of cohort fractions

each lane blocked scenario. These may then be used in conjunction with the percentage of incidents for which the lane groups are blocked (Fig. 4.6) to arrive at the average delay per lane blocking incident and shoulder accident. Building on the previous example, a worksheet example for this calculation is shown in Fig. 4.13. The enclosed CD contains the worksheet file Average delay resulting from incident with the parameters of Fig. 4.13 for the three-lane case.

Each of the four rows for each set of computations provides the delay computation for a cohort. The worksheet columns are described as follows. The data to be entered by the analyst are indicated by an asterisk.

- Column B* – The incident duration from inception to clearance. An incident duration of approximately 53 minutes (0.88 hour) was used for this example.
- Column C* – Roadway capacity in one direction is 6,300 vph.
- Column D* – Demand volume fraction is the representative q/C associated with the cohort. These were obtained from Table 4.2.
- Column E – Demand volume is the product of roadway capacity (Column C) and demand volume fraction (Column D).

AVERAGE DELAY RESULTING FROM INCIDENTS - 3 LANE ROADWAY											
	Data Entry Required										
Section		I-490	Section 61								
A	B	C	D	E	F	G	H	I	J	K	L
	Duration Hr	Capacity	Demand Vol Fraction	Demand Vol	Inc Cap Fraction	Inc. Cap	Delay per incident Veh hr	Frac in Cohort	Del per inc. for coh fraction	Lane weighting for this type	Weighted Delay for Incidents Veh hr
3 Lanes 1 Lane Blocked											
	0.88	6300	0.8	5040	0.49	3087	1928	0.263	507.1		
	0.88	6300	0.6	3780	0.49	3087	342	0.039	13.3		
	0.88	6300	0.33	2079	0.49	3087	0	0.623	0.0		
	0.88	6300	0.2	1260	0.49	3087	0	0.075	0.0		
							Total weighted delay per inc. type		520.5	0.82	427
3 Lanes - 2 Lanes Blocked											
	0.88	6300	0.8	5040	0.17	1071	6378	0.263	1677.3		
	0.88	6300	0.6	3780	0.17	1071	2177	0.039	84.9		
	0.88	6300	0.33	2079	0.17	1071	484	0.623	301.2		
	0.88	6300	0.2	1260	0.17	1071	76	0.075	5.7		
							Total weighted delay per inc. type		2069.1	0.14	290
3 Lanes - 3 Lanes Blocked											
	0.88	6300	0.8	5040	0	0	9757	0.263	2566.2		
	0.88	6300	0.6	3780	0	0	3659	0.039	142.7		
	0.88	6300	0.33	2079	0	0	1201	0.623	748.5		
	0.88	6300	0.2	1260	0	0	610	0.075	45.7		
							Total weighted delay per inc. type		3503.2	0.04	140
							Delay for average incident				857

Fig. 4.13 Worksheet for calculation of average delay resulting from incident

- Column F*– The incident capacity fractions were obtained from Table 4.1.
- Column G – The incident capacity is the product of the roadway capacity (Column C) and incident capacity fraction (Column F).
- Column H – The delay per incident is provided by the application of (4.4) to the data in the previous columns.
- Column I* – The fraction of the traffic in the cohort is obtained from the cohort fraction analysis (Fig. 4.12).
- Column J – The delay per incident for the cohort fraction is the product of the delay per incident (Column H and the fraction of the traffic in the cohort (Column I). These are then summed over all cohorts for the type of incident to provide total weighted delay for the incident type.
- Column K – The lane weighting represents the fraction of lane blocking incidents for the data set. It was obtained from Fig. 4.6 and must be entered by the user if changes are required.
- Column L – The weighted delay for incidents is the product of the delay per incident for the cohort fraction (Column J) and the lane weighting (Column K).

The delay for the average incident is the sum of the weighted delays for each incident type.

The saving in delay due to ITS measures may be computed by taking the difference in the worksheet computation for which incident duration (Column B) is entered prior to ITS project implementation and after its implementation. The Design ITS model's default values for time saving per incident (TSI) are shown in Table 4.3. They are based on a five minute saving in the time to clear the incident resulting from the ITS technologies and procedures described in Sect. 4.5. This figure is the representative for agencies that implement ITS technologies intensively and use their resources effectively. The effectiveness of agency use of ITS is discussed in the Traffic Incident Management Self Assessment Report [9].

These factors are used in the benefits evaluation model described in Sect. 4.6.3.2.

Table 4.3 Representative values of time saved per incident (TSI) for intensive ITS implementations

Number of lanes	Traffic condition level (Sect. 4.5.1.2)	Vehicle time saved per incident (vehicle hours)
2	1	27.7
2	2	92
2	3	183.8
3	1	39.4
3	2	83.3
3	3	271.8
4	1	31.3
4	2	71.3
4	3	254.9

4.5 Functional Requirements for Improving Incident Response and Relationship of Improvement Techniques

From a traffic management center perspective, incident management may be described by the steps shown in Fig. 4.1 and discussed in Sect. 4.1.

Incident management is a multiagency function involving a number of emergency responders. ITS and traffic management centers can assist in incident management by helping to plan the management of an incident, by providing surveillance and monitoring capability and by facilitating communication among incident responders. Table 4.4 describes a number of functions that can be provided by ITS and traffic management centers.

The relationship between incident management functions and a number of traffic operations and ITS technologies is shown in Table 4.5. References [1, 7] discuss

Table 4.4 ITS and TMC related incident management functions

	Stages of incident management			
Incident management function	Detection and verification	Response	Clearance	Recovery
Coordinate development, archiving and update of incident response plans	√	√	√	√
Integrate traffic incident management needs into operations planning and ITS design	√	√	√	√
Detect an incident and identify its properties (lanes blocked, location)	√			
Classify incident into type and severity	√			
Notify responding agencies of location and character of incident	√			
Improve conventional traffic operations		√	√	
Select response plan. Implement: • Lane and ramp controls • Emergency signal timing plans • Motorist information		√	√	√
Facilitate operations for emergency responders		√	√	√
Provide traffic conditions and/or route guidance to emergency responders		√	√	√
Provide incident information to motorists (see Chap. 5)		√	√	√
Provide tail of queue detection		√	√	√
Terminate incident management functions				√

Table 4.5 Traffic operations and ITS technologies for incident response

Operations and technologies	Functions						
	Incident detection and verification	Incident classification	Response plan implementation	Emergency responder operations	Traffic information for emergency responders	Tail of queue detection	Termination of incident management
CCTV	√	√	√	√	√	√	√
Traffic detectors	√		√		√	√	√
911 service	√	√					
Traffic reporting service	√		√		√		
Service patrols	√	√		√			
Police services	√	√	√	√			
TMC information displays	√	√	√				
Lane controls			√				
Ramp meters			√				
Emergency signal timing plans			√				
Incident command system		√	√	√			
Surveillance and travel time on alternate routes			√	√	√		
Traffic information for emergency service providers				√	√		
In vehicle information for emergency service providers				√	√		

these operations and technologies. Sections 4.5.1 and 4.5.2 discuss the effective application of these operations and technologies except for motorist information which is discussed in Chap. 5.

4.5.1 *Improving Incident Detection and Verification*

The following sections discuss commonly used approaches for incident detection and verification.

4.5.1.1 Public Service Access Points

Most freeway incidents are detected by motorist cellular telephone calls to public service access points (PSAPs) providing 911service. In many cases, this information is provided to traffic management centers on a data channel using filters to restrict the information to traffic related incidents. When the call originates from a vehicle involved in an incident, position information is automatically available. In some cases, the information provided by the call is suitable for verification, but in other cases, other verification techniques (usually CCTV observation, police responses, or motorist service patrols) are required.

4.5.1.2 CCTV

CCTV and traffic detectors may detect incidents more rapidly than other measures such as 911 calls. CCTV provides the capability to identify the type of incident and its properties, thus providing verification as well as detection. This information may be used to determine the types of emergency services required. Verification by CCTV is often the fastest way to accomplish this function. In many cases, the number of CCTV cameras on the freeway far exceeds the number of monitors available in the traffic management center. When this occurs, automated camera "tours" are commonly employed to allow the operator to rapidly monitor large sections of the roadway.

Conventional CCTV cameras require the roadway to be lighted in order to fully support incident management operations. It is possible to couple conventional cameras with infrared cameras to extend the capability for observation when roadways are unlit.

Incidents that are not the result of accidents are generally distributed uniformly along the roadway. Accidents more frequently occur within or near interchanges. Furthermore, accidents generally require more time to clear. Thus, the benefits for cameras are greatest when they are placed in the interchange area. Figure 4.14 shows the general relationship for cost and benefits of CCTV placement. Curve A represents the benefit as a function of percent coverage when demand volumes are

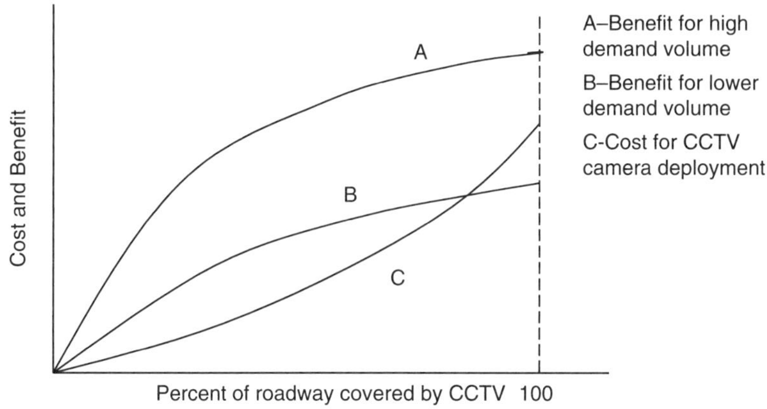

Fig. 4.14 Benefit and cost of CCTV coverage

Table 4.6 Considerations for preliminary location of CCTV cameras

Deployment level	Application factor
1	High accident rate locations. Other locations as deemed necessary
2	High accident rate locations, freeway-to-freeway interchanges, spot congestion locations. Other locations as deemed necessary
3	Continuous coverage (average 1.5–2.0 cameras per centerline mile)

high. The concave shape results from the higher benefits of the cameras at or near the interchange area where the accident rate is generally higher than in the roadway portion away from the interchange. Curve B shows the benefits when the demand volume is lower (see Fig. 4.4 as an example of incident delay sensitivity). Curve C is the cost of installing CCTV cameras. Curve C is convex because as the percentage of the roadway covered increases, the likelihood of overlapping camera coverage increases. This increases the number of cameras employed per mile of roadway. For each deployment alternative (represented by a value for the roadway percent covered), the line segment between the benefits curve and the cost curve represents the net benefits. The net benefit can possibly become negative when excessive CCTV coverage is provided for low demand volume situations.

Table 4.6 provides guidance for the initial consideration of camera locations which are defined in terms of deployment "levels" [8, 10] (see Chap. 3) as follows:

- Level 3 – A continuous section of roadway in one direction that includes peak hour level of service D or worse traffic for at least one half of the section.
- Level 2 – A continuous section of roadway in one direction that includes peak hour level of service C or worse traffic for at least one half of the section. Conditions worse than level of service C may be present at scattered spot locations or for small sections. In this case, it may be appropriate to increase the concentration of field equipment at these locations.
- Level 1 – Traffic conditions better than level 2.

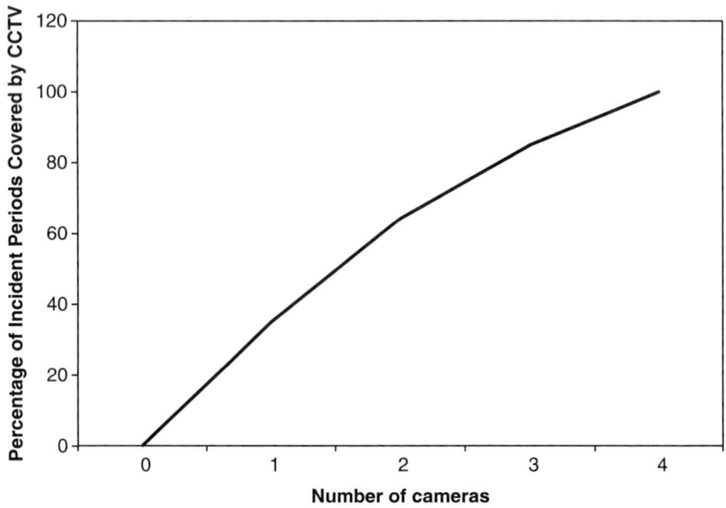

Fig. 4.15 Incident periods covered vs. number of cameras

Because deployment levels and application factors cover broad ranges of conditions, the consideration of CCTV deployment alternatives is recommended.

Appendix B describes a measure for the relative effectiveness (RTV) of CCTV coverage based on the fraction of incident periods observable by CCTV. An example of this measure for a particular roadway section is shown in Fig. 4.15. A worksheet (*RTV*) for computing RTV from section accident data is described in Appendix B and is provided on the CD.

4.5.1.3 Traffic Detectors

Point and probe detectors are discussed in the following sections.

Point Detectors

Point detectors sense lane volume, speed, and lane occupancy at a particular location. All detector types do not necessarily sense all variables. Most point detectors can provide data for each traffic lane. Aggregation of detector data into the traffic parameters is performed periodically.

Detector technologies that are commonly used for freeway monitoring include:

- Inductive loop detectors. While probably the most accurate and frequently selected detector type, the difficulty of maintaining inductive loop detectors over long periods of time has led some agencies to use other technologies.
- Frequency modulated continuous wave radar detectors. This type of detector has found increasing use because of its adequate performance and low maintenance requirement.

- Video processing detectors. The need to readjust these detectors from time to time as well as lighting, sun glint, and weather considerations has encouraged some agencies to select alternative technologies for freeway applications.
- Acoustic detectors.

Klein [11] discusses detector technologies in detail. Detector data is most commonly displayed in the traffic management center as a map presentation that is color coded according to speed. Closer detector spacing improves the location of the lower speed indications that may result from an incident. Using historical experience, operators often infer the presence of a potential incident, and use other means such as CCTV, 911 information, service patrols, and police reports to confirm the incident and to provide additional information. Point detectors also provide information for other than incident detection functions as described in Chaps. 5–7. Reference [10] provides guidance for point detector spacing for traffic condition level 3 (defined in Sect. 4.5.1.2.) For incident detection and general surveillance applications, detector station spacing should not exceed 0.3–0.4 miles (0.48–0.64 km) for this traffic condition level. A general approach for placing detectors is

- Identify sites for mainline stations required for ramp meters if they are to be employed.
- Add additional stations so that each mainline roadway portion between each entry and exit location has a station.
- Add detectors so that gaps do not exceed 0.33 miles (0.53 km).
- Modify detector locations to avoid very short spacing.

Typical spacing on freeways with ramp metering may average 0.25–0.33 directional miles (0.40–0.53 directional kilometers).

Point detectors may be used for incident detection as described in the following sections.

Point Detectors Without Incident-Detection Algorithms

Many agencies use point detectors to help assist in incident detection. For example, map displays show the speed for zones in the roadway. The zone is usually defined as the approximate midpoint between the locations of two detector stations. While the display does not directly indicate an incident in the zone, an experienced operator, using his historical knowledge of expected traffic conditions, may often infer the possible presence of an incident, particularly when the incident occurs in off-peak periods. This tentative detection requires confirmation by other means.

Traffic flow at any point on the roadway is governed by the general traffic equation discussed in Sect. 3.1.1.1 and is repeated below

$$q = ku \tag{4.8}$$

where

q = volume (vehicles per hour per lane)
k = density or concentration (vehicles per mile per lane)
u = space mean speed (miles per hour)

When a traffic incident occurs that results in the demand upstream of the incident exceeding the remaining capacity of the roadway, a queue begins to form at the incident site, thus increasing density and reducing the speed upstream of the incident. The queue builds with time and causes a shock wave to propagate upstream of the incident. The time required for this shock wave to reach the traffic detector determines, in large measure, the time delay in the operator's ability to detect the incident in this way. The following discussion provides an example of how this time may be estimated. The equation for the velocity of propagation of the shock wave is [12]

$$w_{AB} = (q_A - q_B)/(k_A - k_B) \tag{4.9}$$

where k and q are as defined above. Subscript A identifies the conditions in the queue downstream of the shock wave boundary, and subscript B describes the conditions upstream of the boundary.

The following is an example of the average incident detection time for point detectors when applied in this way. Figure 4.16 diagrams the propagation of the shock wave and describes the conditions found in the example.

Using the following parameters, the density upstream and downstream of the incident is

$q_A = 1{,}100$ vehicles per hour
$q_B = 1{,}600$ vehicles per hour
$k_A = 110$ vehicles per mile
$k_B = 26.7$ vehicles per mile

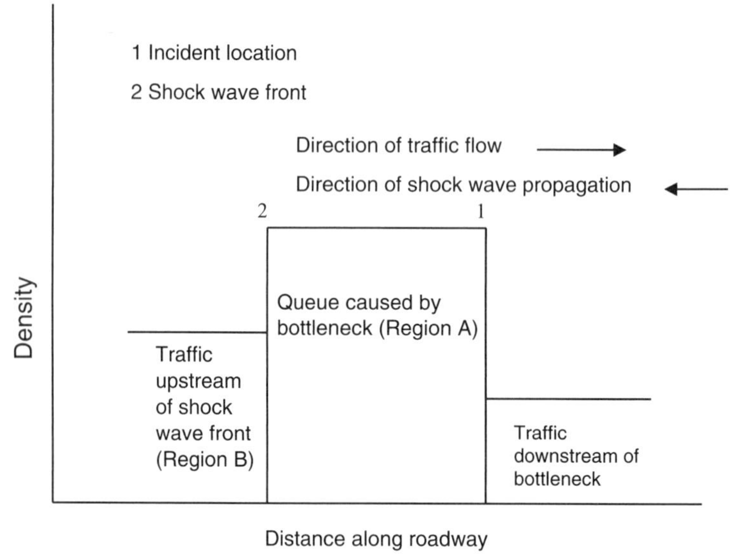

Fig. 4.16 Shock wave propagation for a bottleneck caused by an incident

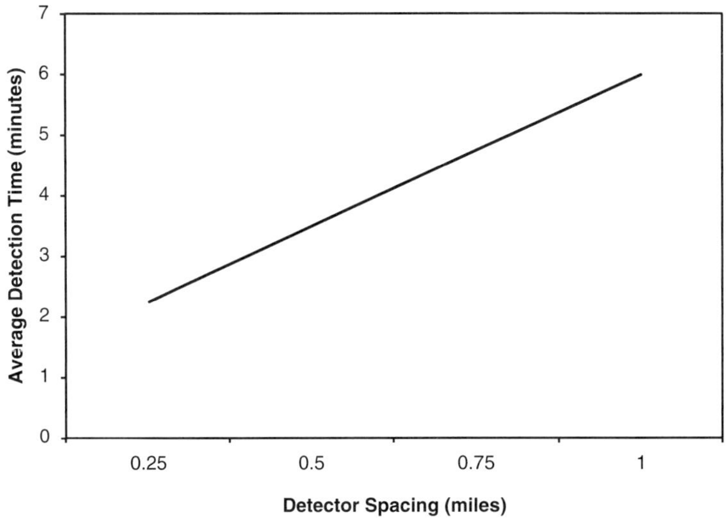

Fig. 4.17 Example of average detection time

Applying (4.9) a shock wave propagates upstream of the incident at a speed of 6 mph. If detector stations are spaced at intervals of 0.25 miles, the wave will take 2.5 min to travel that distance (or an average propagation time of 1.25 min assuming that incidents occur randomly spaced between detectors will result.) To this must be added the time to smooth and process the detector data (typically a minimum of one minute). Figure 4.17 shows the average detection times for different detector spacing based on this example.

Since detection time depends on the incident scenario, it is useful to define a parameter B1 to represent the timely detection probability for a range of scenarios for different detector spacings. The models for incident management effectiveness discussed in Sect. 4.6.3 employ parameter B1. An example of a value set for B1 is provided by the following expression based on average point detector spacing [13]:

0.4 mi ≥ spacing, B1 = 0.9
0.7 mi ≥ spacing > 0.4 mi, B1 = 0.7
1.0 mi ≥ spacing > 0.7 mi, B1 = 0.5
2.0 mi ≥ spacing > 1.0 mi, B1 = 0.3
Spacing > 2 mi, B1 = 0.1
No detectors, B1 = 0.0

Point Detectors with Automatic Incident-Detection (AID) Algorithms

The previous section describes a manual technique for incident detection. Although not employed by most agencies, techniques that process point detector data have been used, with varying degrees of success, to alert the operator to the possible presence of an incident [4, 14, 15]. Criteria for assessing the quality of AID algorithms include detection rate, false alarm rate, and time to detect. On the basis of these criteria, the performance of these algorithms has generally not measured up to that achievable by other means. Martin [14] concludes that "the use of video coverage on the freeway systems and ever-expanding use of cell phones have made the need for AID algorithms less important."

Probe Detection

Probe detection consists of determining the difference in arrival times of a vehicle at two preestablished locations. Some technologies use roadway equipment to establish these locations, other technologies provide virtual locations. Common probe detection technologies include:

- Toll tags and toll tag readers
- Cellular telephone based detection
- License plate readers
- GPS based probe services
- The Vehicle Infrastructure Integration (VII) initiative will, in time, provide probe based information

While probe detection is most commonly used to display travel times or for general traffic surveillance applications it may be used to provide incident detection capability. An example of the use of probe detection for incidents is provided by Mouskos [16].

The Transmit system is a probe system based on the use of toll tag readers. Data from a set of readers deployed on the New York State Thruway and on the Garden State Parkway in New Jersey are processed in the following way for the purpose of detecting incidents. A historical database of expected travel times for 15-min periods is established and updated. A false alarm probability is computed for each vehicle based on the measured difference between normal and observed travel times. The vehicle is considered to be late due to an incident when the travel time exceeds three times the standard deviation of the normal travel time. Other factors were also taken into consideration.

Test results showed the probability of incident detection to be approximately 89% on the New York State Thruway and 72% on the Garden State Parkway. False alarm probabilities ranged from 10 to 22% on the New York State Thruway and from 16 to 32% on the Garden State Parkway. In recent years, the Transmit system has been extended in the New York City metropolitan area and to other locations on the New York State Thruway [16].

Incident detection time, and the precision with which an incident may be located, depends on the distance between the locations used to estimate the travel times. In the case of infrastructure based readers, reducing this distance requires significant expenditure.

4.5.1.4 Motorist Service Patrols and Police

Motorist service patrols are available in many locations during significant portions of the work-week. They may be available at other times as well. Motorist service patrols not only detect and confirm incidents but also help clear minor incidents and assist other responders in incident clearance and traffic management.

Police patrols also detect and confirm accidents. Police are often the first responders when incidents have been detected.

Traffic Management Center Support

In addition to coordinating the detection of incidents as described above, traffic management centers (TMCs) classify incidents and assist emergency responders in their clearance. Software tools are available to assist in this process. Figure 4.18 [17] is an example of an incident management screen found in New York State to log incident information.

4.5.2 Improving Incident Response, Clearance, and Recovery Through ITS

Several techniques utilized by TMCs to assist emergency responders in incident clearance are discussed in the following sections.

4.5.2.1 Incident Response Plans

The coordination of incident response planning is generally performed by the stakeholders, often through a group or committee established for that purpose. The traffic management center's responsibilities typically include:

- Development of a coordinated set of motorist information messages.
- Identification of suitable alternate routes and development of emergency signal timing plans to support diverted traffic.
- Operation of ancillary controls and displays such as lane control signals, blank out signs, and dynamic trail-blazers.

Incidents are often classified by severity, and response plans implemented on that basis. Figure 4.19 [18] shows an example of a classification scheme and its relationship to the selection of emergency plans.

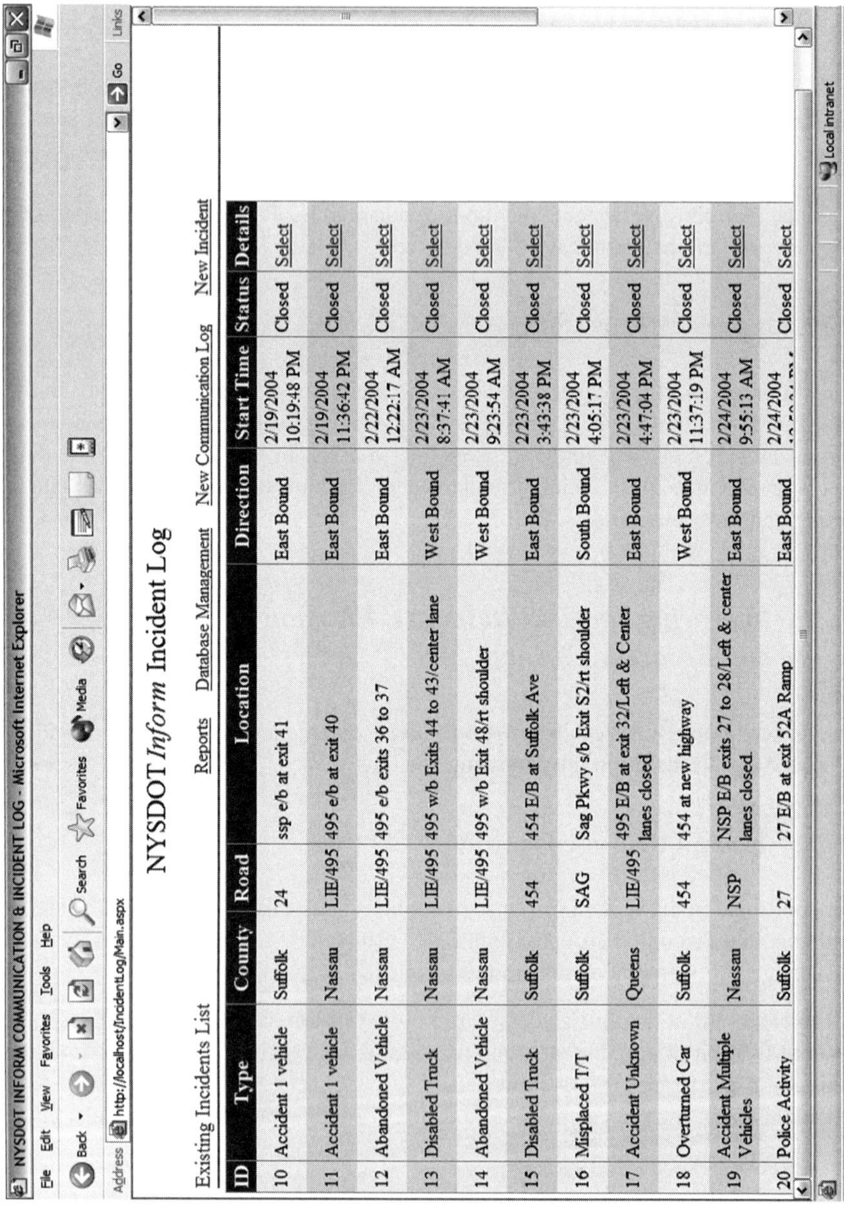

ID	Type	County	Road	Location	Direction	Start Time	Status	Details
10	Accident 1 vehicle	Suffolk	24	ssp e/b at exit 41	East Bound	2/19/2004 10:19:48 PM	Closed	Select
11	Accident 1 vehicle	Nassau	LIE/495	495 e/b at exit 40	East Bound	2/19/2004 11:36:42 PM	Closed	Select
12	Abandoned Vehicle	Nassau	LIE/495	495 e/b exits 36 to 37	East Bound	2/22/2004 12:22:17 AM	Closed	Select
13	Disabled Truck	Nassau	LIE/495	495 w/b Exits 44 to 43/center lane	West Bound	2/23/2004 8:37:41 AM	Closed	Select
14	Abandoned Vehicle	Nassau	LIE/495	495 w/b Exit 48/rt shoulder	West Bound	2/23/2004 9:23:54 AM	Closed	Select
15	Disabled Truck	Suffolk	454	454 E/B at Suffolk Ave	East Bound	2/23/2004 3:43:38 PM	Closed	Select
16	Misplaced T/T	Suffolk	SAG	Sag Pkwy s/b Exit S2/rt shoulder	South Bound	2/23/2004 4:05:17 PM	Closed	Select
17	Accident Unknown	Queens	LIE/495	495 E/B at exit 32/Left & Center lanes closed	East Bound	2/23/2004 4:47:04 PM	Closed	Select
18	Overturned Car	Suffolk	454	454 at new highway	West Bound	2/23/2004 11:37:19 PM	Closed	Select
19	Accident Multiple Vehicles	Nassau	NSP	NSP E/B exits 27 to 28/Left & center lanes closed	East Bound	2/24/2004 9:55:13 AM	Closed	Select
20	Police Activity	Suffolk	27	27 E/B at exit 52A Ramp	East Bound	2/24/2004	Closed	Select

Fig. 4.18 Sample incident logging screen from freeway management software used in New York State

		Lanes Impacted / Action Level			
Time of Day	**Estimated Duration**	0 Lanes	1 Lane	2 Lanes	>2 Lanes
0000 - 0600	< 2 hours	0	0	1*	3*
	2-4 hours	0	0	2*	3*
	>4 hours	0	0	2*	3*
0600 - 1000	< 0.5 hours	1	1	2	3
	0.5 - 2 hours	1	1	2	4
	> 2 hours	1	2	3	4
1000 - 1500	< 2 hours	1	1	2	3
	2-4 hours	1	1	2	3
	> 4 hours	1	2	3	3
1500 - 1900	< 0.5 hours	1	1	2	3
	0.5 - 2 hours	1	1	2	4
	> 2 hours	1	2	3	4
1900 - 2400	< 2 hours	0	0	1*	3*
	2-4 hours	0	0	2*	3*
	> 4 hours	0	0	2*	3*

Level 0 No special action required

Level 1 Implement Response Plan to notify appropriate PSAs
Turn on Level 1 CMS and HAR

Level 2 Implement Response Plan to notify appropriate PSAs
Turn on Level 2 CMS and HAR
Turn HAR flashing lights on a level 2

Level 3 Implement Response Plan to notify appropriate PSAs
Turn on Level 3 CMS and HAR
Turn HAR flashing lights on a level 3
Provide Advisory Alternate Routing

Level 4 Implement Response Plan to notify appropriate PSAs
Turn on Level 4 (and above) CMS and HAR
Turn HAR flashing lights on a level 4
Provide Mandatory Alternate Routing

Level n CMS n = number of decision points prior to the incident corridor
Level n HAR n = number of times the related advisory is repeated in a HAR cycle (e.g. within a 3 minute cycle)
Level n* * = notification of operations personnel may be required to implement outside normal duty hours

Fig. 4.19 ARTEMIS incident classification matrix

4.5.2.2 Improved Interagency Communication

Improved sharing of information and coordination of response operations may be facilitated by the following:

- Collocation of management centers – Those which are most commonly collocated include operation centers for freeway traffic management, traffic signal control, police operations, and emergency management centers.

- Use of ITS standard protocols – These protocols facilitate the migration of information among management centers. Management centers commonly support the National Transportation Communications for ITS Protocol (NTCIP) [19] and the IEEE 1512 standards suites [20].
- Special systems for interagency communications – In recent years, some regions have begun using special systems for interagency communications. These systems emphasize two-way communication with emergency responders. For example, an interagency coalition in New York City deployed a major incident management system, known as the Integrated Incident Management System (IIMS) [21]. IIMS implements a partnership between New York State DOT, New York City Police Department, and the New York City Department of Emergency Management. Figure 4.20 shows an example of one type of display that is provided to the vehicle. The responders in the field may enhance this information or provide additional photos to supplement the report.

4.5.2.3 Provision of Traffic Information to Responders

This information may provide speed, travel time, or routing. If the information for the freeways and major surface streets is provided by TMCs to emergency service providers, they often have the capability to alter their access routes to the incident, or to use vehicles located more favorably with respect to congestion patterns. This information is best provided in the form of speed or congestion maps, or by instructions depicting the quickest route. Updated traffic information may be provided to

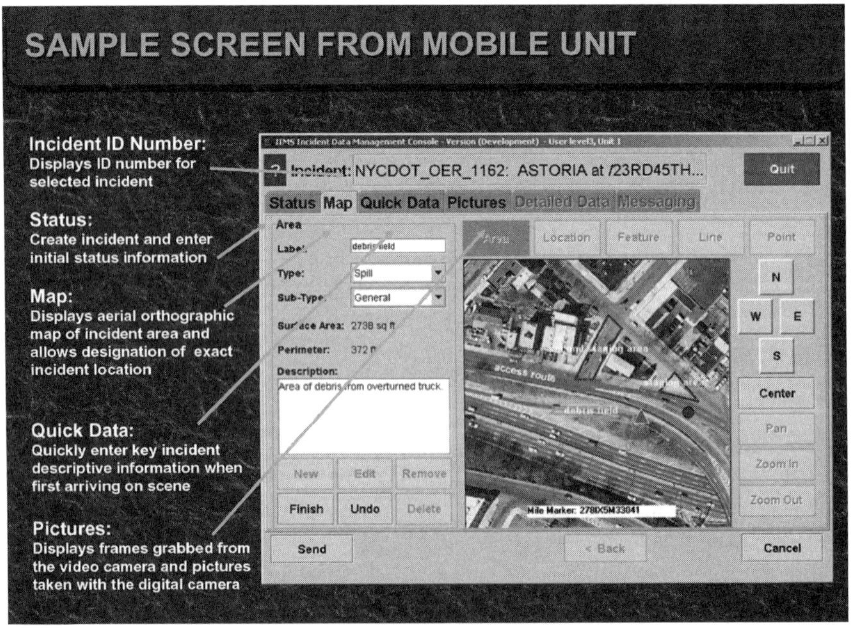

Fig. 4.20 Example of IIMS in-vehicle display [21]

drivers of emergency response vehicles by means of in-vehicle displays. Provision of this capability requires knowledge of travel times on the freeway and on the alternate routes that may be employed to access the incident. Table 4.7 identifies the mature and emerging technologies that provide this information.

4.5.2.4 Definition and Management of Major Response Routes

The following emergency route management measures will facilitate more rapid access.

- Traffic signal preemption – Techniques include preemption of individual traffic signals by the use of wireless, optical, or sonic communication links to the intersection. Traffic signals may also be preempted along preplanned routes from designated dispatch centers such as fire department facilities.
- Establishment of priorities for roadway maintenance and construction operations that benefit emergency response vehicles – Lane closures or other capacity reducing maintenance activities along routes commonly used by emergency responders may increase travel time on these routes. Improved emergency vehicle access is facilitated by planning maintenance activities so that lane closure time is minimized, and by coordinating maintenance and construction so that alternate routes used by emergency vehicles are not simultaneously impacted. Similarly, providing priority treatment for snow and ice removal along these routes will improve response time. Input from emergency service responders can assist in identifying specific locations for priority treatment [22].

Table 4.7 Technologies for travel time determination [22]

	Freeways	Surface Streets
Mature technologies	Point detectors (intensively deployed)	Point detectors associated with adaptive traffic signal control systems. Additional detectors may be employed to support this function.
	Probe detection based on electronic toll collection or license plate reader technology	Probe detection based on electronic toll collection or license plate reader technology
Emerging technologies	GPS probe based traffic surveillance	GPS probe based traffic surveillance
	Cellular telephone based probe traffic surveillance	Cellular telephone based probe traffic surveillance
	Probe techniques based on vehicle infrastructure initiative (IntelliGuide)[a] technology	Probe techniques based on vehicle infrastructure initiative (IntelliGuide)[a] technology
		Analysis models used in conjunction with point detector measurements

[a] IntelliGuide was formarly known as the Vehicle Infrastructure Initiative (VII)

- Emergency vehicle turnarounds on freeways – Travel times from emergency vehicle locations to high incident rate freeway sections may be improved by using appropriately located emergency vehicle turnarounds. Access to these turnarounds may be protected, if necessary, by the use of gates that are remotely operated by emergency vehicles.
- Coordination of traffic calming plans with emergency vehicle route requirements – Traffic calming treatments often have an adverse effect on the speed and accessibility of emergency service vehicles. It is important for the agencies involved in developing traffic calming plans to coordinate with emergency service providers. Atkins and Coleman [23] discuss the effect of traffic calming measures on emergency vehicle speeds. This research has led Portland, Oregon to classify a number of streets as major emergency response streets. Traffic preemption devices are emphasized, and traffic-slowing devices are avoided on these streets. Other agencies [24] have also addressed this issue.

4.5.2.5 Tail of Queue Detection

During the progress of an incident, the tail of the queue progresses upstream. The tail may be at a considerable distance from the incident site, making its location difficult to identify by on-site personnel. Point detectors and CCTV may be used to provide this information (CCTV provides a labor intensive solution that may be inappropriate when a number of incidents must be simultaneously monitored.) Detecting the tail of the queue serves the following functions:

- Assists on-site personnel in adopting appropriate traffic management measures.
- Assists emergency vehicles in finding the best route to the incident.
- Assists in the selection of appropriate motorist information messages and routing plans. As the tail of the queue continues to build after the incident has been cleared, this function continues, even after the emergency responders have left the scene.
- Termination of queue. It is important to detect this event so that motorist information messages do not indicate the presence of an incident after the queue is cleared. Point detectors and CCTV may be employed for this purpose.

4.6 Measuring Incident Management Effectiveness

Incident management effectiveness may be viewed from a number of perspectives as described in the following sections.

4.6.1 Degree of Attainment for Recommended Management Functions, Operations, and Technologies

FHWA has established a Traffic Incident Management Self-Assessment program for this purpose. An example of the type of issues covered and the score from responders is shown in Table 4.8.

Table 4.8 Example of self-assessment questions and responses [25]

Question number	Question	Percent of assessments scoring 3[a] or higher
4.2.1.1.	Have established criteria for what is a "major incident" – incident levels or codes?	17%
4.2.1.2.	Identify high ranking agency members available on 24/7 basis to respond to a major incident?	77%
4.2.1.3.	Have a preidentified (approved) contact list of resources (including special equipment) for incident clearance and hazardous materials response?	66%
4.2.1.4.	Have the response equipment prestaged for timely response?	44%
4.2.2.1.	Train all responders in traffic control procedures?	30%
4.2.2.2.	Utilize on-scene traffic control procedures for various levels of incidents in compliance with MUTCD?	29%
4.2.2.3.	Utilize traffic control procedures for the end of the incident traffic queue?	14%
4.2.2.4.	Have mutually understood equipment staging and emergency lighting procedures on-site to maximize traffic flow past an incident while providing responder safety?	16%
4.2.3.1.	Utilize the Incident Command System?	54%
4.2.3.2.	Have specific policies and procedures for fatal accident investigation?	51%
4.2.3.3.	Have specific policies and procedures for hazardous materials response?	69%
4.2.3.4.	Have quick clearance policies?	36%
4.2.3.5.	Have a prequalified list of available and contracted towing and recovery operators (to include operators' capabilities)?	74%
4.2.3.6.	Use motorist assist service patrols?	70%

[a] The highest score is 5

4.6.2 General Measures

General quantitative measures for the assessment of incident management that have been proposed by the National Transportation Operations Coalition, among others, are shown in Table 4.9.

4.6.3 Model for Evaluating Incident Management Effectiveness

The following sections discuss a model for evaluating alternative ITS design concepts as applied to incident management. The model computes a parameter (H) ranging from 0.0 to 1.0 that represents the potential ability of the ITS to effectively

Table 4.9 NTOC proposed performance measures for incident related travel delay [9]

Measure	Definition	Sample units of measurements
Incident duration	The time elapsed from the notification of an incident until all evidence of the incident has been removed from the incident scene	Median minutes per incident
Nonrecurring delay	Vehicle delays in excess of the recurring delay for the current time-of-day, day-of-the-week, and day type	Vehicle-hours
Travel time reliability (buffer time)	The buffer time is the additional time that must be added to a trip to ensure that 95% of the travelers making the trip will arrive at their destination before the intended time	This measure, delay in minutes, may also be expressed as a percent of total trip time or as an index

provide incident management support. The model is implemented in the Design ITS software model [13].

4.6.3.1 Relationship of Incident Management Functions to Techniques and Technologies

The parameter H incorporates the following functional requirements:

1. Timely detection of incidents (H1)
2. Timely confirmation and classification of incidents (H2)
3. Timely assistance provided by the TMC in managing incident response and clearance (H3)
4. Timely detection of the tail of the queue and the termination of the queue (H4)

The probability that each of the above events has the potential to be satisfied depends on the technologies and operations employed.

Table 4.10 uses the index g to represent these functional requirements. V_{ng} is a value that represents the potential ability of the technology or operation to satisfy requirement g. Representative values are provided in Table 4.11.

Default probabilities provided in the Design ITS model for V_{ng} are shown in Table 4.11 [13].

The user may choose to modify these values based on experience or based on special considerations pertaining to the application.

4.6.3.2 Model for Incident Management Effectiveness Potential

A value H_g is required to represent the effectiveness potential for each of the incident management functional requirements. Since the contributions of the technologies

Table 4.10 Probability of satisfying functional requirement

Technologies and operations	Probability that requirement is satisfied
1. 911/PSAP information availability	V_{1g}
2. Police operations	V_{2g}
3. CCTV	V_{3g} • RTV
4. Motorist service patrols	V_{4g} • K35 • K40
5. Electronic traffic detection (point or probe)	V_{5g} • B1

In the table, B1 = timely detection probability for a range of scenarios for different detector spacing (Sect. 4.5.1.3); RTV = fraction of roadway incidents observed by CCTV (Appendix B); K35 = correction factor for level of service (Appendix F); K40 = scaling factor for service patrols (Appendix F)

Table 4.11 Default values for V_{ng}

	Technologies				
Functional requirements (g)	V_{1g}	V_{2g}	V_{3g}	V_{4g}	V_{5g}
1. Timely detection	0.6	0.3	0.9	0.5	0.4
2. Timely conf. & class	0.3	0.6	0.9	0.5	0.2
3. Timely assistance	0.0	0.9	0.8	0.5	0.2
4. Tail of queue and termination	0.0	0.1	0.5	0.2	0.8

toward supporting incident management overlap, a technique is required to model these interactions. The technique employed uses a combination of probability theory and Bayesian inference (for example see Klein [11]). To compute H_g

$H_g = 1 -$ probability that no technology satisfies function g

Probability that no technology satisfies the function g = (1 − probability of satisfaction by technology 1) • ....• (1 − probability of satisfaction by technology 5).

Thus, H_g is provided by

$$H_g = 1 - (1 - V_{1g}) \bullet (1 - V_{2g}) \bullet (1 - V_{3g} \bullet RTV) \bullet (1 - V_{4g} \bullet K40 \bullet K35) \bullet (1 - V_{5g} \bullet B1) \quad (4.10)$$

Incident management effectiveness potential includes the following operations:

- Timely detection of incidents (H_1)
- Timely confirmation and classification of incidents (H_2)
- Management of incidents (H_M)
- Timely assistance to emergency responders (H_3)
- Timely detection of tail of queue and termination of queue (H_4)

Since H_3 and H_4 contribute to incident management, a measure of H_M may be provided by

$$H_M = Y \bullet H_3 + (1 - Y) \bullet H_4 \quad (4.11)$$

where Y is the fraction of the incident management operation represented by the timely assistance to emergency responder function. The Design ITS model includes a default value of 0.8 for Y.

The value of H is provided by

$$H = H1 \bullet H2 \bullet H_M \tag{4.12}$$

4.6.3.3 Benefits for ITS Incident Management Treatment

Reduction in incident clearance time produces the following benefits:

- Reduction in overall vehicle delay
- Reduction in secondary accidents as a result of reduction in the time that the queue is present
- Reduction in fuel consumption
- Reduction in vehicle emissions

A model for reduction in delay and accidents is shown in Fig. 4.21.

Representative values for specific parameters in the model are shown in Table 4.12.

Fuel reduction is modeled as a multiplier of the delay reduction. It is estimated by the expression:

$$FS = ((1 - CVF) \bullet K21 + CVF \bullet K22) \bullet TS \tag{4.15}$$

where

FS = Annual fuel reduced (gallons)

CVF = fraction of traffic consisting of commercial vehicles

K21 = Auto fuel consumption rate in congestion (gallons/hour). A typical value is 1.05 gal/h [13].

K22 = Commercial vehicle fuel consumption rate in congestion (gallons/hour). A typical value is 1.97 gal/h [13].

The emissions reduced are modeled as a multiplier of the delay reduced. It is estimated by the expression:

$$ER = ((1 - CVF) \bullet K3 + CVF \bullet K4 \bullet TS \tag{4.16}$$

where

ER = Emissions rate in grams/hour

K3 = Auto emissions rate in congestion (grams/hour)

K4 = Commercial vehicle emissions rate in congestion (grams/hour)

Representative values for K3 and K4 are provided in Table 4.13.

The computations for (4.10) through (4.14) are provided by the *Inc mgt effectiveness potential worksheet* on the enclosed CD. An example of a design tradeoff to obtain benefits for alternative CCTV camera and detector deployments and an illustration of a worksheet is provided in Appendix C.

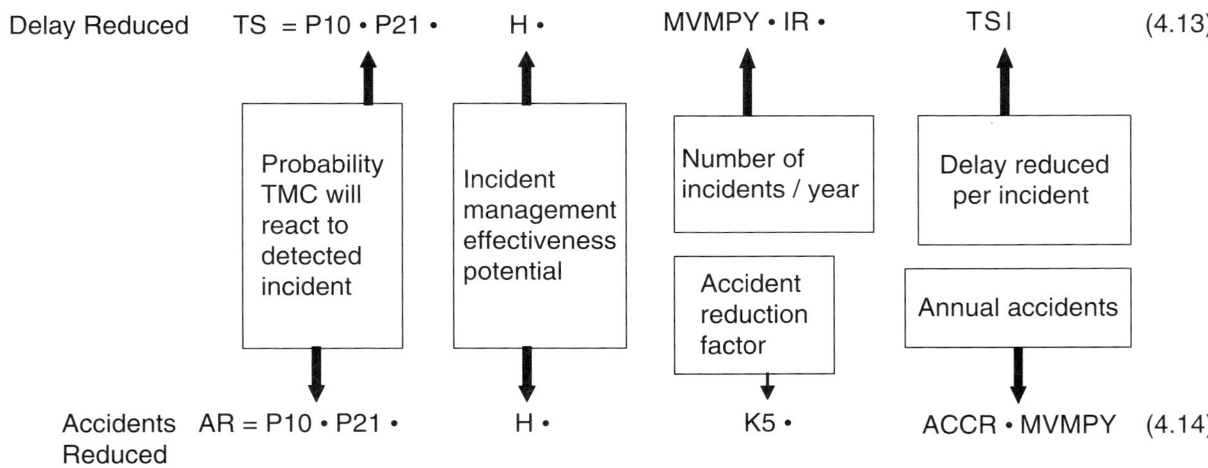

Delay Reduced $TS = P10 \cdot P21 \cdot H \cdot MVMPY \cdot IR \cdot TSI$ (4.13)

Accidents Reduced $AR = P10 \cdot P21 \cdot H \cdot K5 \cdot ACCR \cdot MVMPY$ (4.14)

P10 – Probability incident managed if TMC manned

P21 – Probability TMC manned when incidents occur

H – Incident management effectiveness potential

MVMPY – Millions of vehicle miles per year

IR – Incidents per million vehicle miles

TSI – Time saved per incident

K5 – Accident reduction factor

ACCR – Accident rate

Fig. 4.21 Model for delay and accident reduction benefits

Table 4.12 Representative parameter values for delay and accident reduction benefits model

Symbol	Parameter	Representative value
P10	Probability incident is managed if TMC is staffed	1.0 if TMC operations manual or operating direction requires support of incidents.
P21	Probability TMC staffed when incidents occur	1.0 if TMC operates around the clock. If TMC is staffed on weekdays from 6 AM to 7 PM and with a representative time of day and traffic volume distribution, 58.6% of the vehicle miles traveled and accidents will occur during this period [8].
K5	Accident reduction factor (fraction of accidents reduced by ITS support of incident operations)	0.10 is used as the default factor in the Design ITS model [13].

Table 4.13 Representative emissions rates [13]

	Auto emission factor K3	Commercial vehicle emission factor K4
Hydrocarbons	43.0	75.9
Carbon monoxide	355.5	751.9
Oxides of nitrogen	17.8	142.8

References

1. Neudorff LG et al (2003) Freeway management and operations handbook. Report FHWA-OP-04-003. Federal Highway Administration, Washington, DC
2. Ozbay K, Kachroo P (1999) Incident management in intelligent transportation systems. Artech House, Boston
3. Lindley JA (1987) A methodology for quantifying urban freeway congestion. Transportation Research Record 1132. Transportation Research Board, Washington, DC
4. Raub RA (1997) Secondary crashes: An important component of roadway incident management. Transport Q 51(3):93–104
5. Morales JM (1986) Analytical procedures for estimating freeway traffic congestion. Public Roads 50(2):55–61
6. Presley MW, Wyrosdick KG, Sulbaran TA (2000) Calculating benefits for NAVIGATOR, Georgia ITS. 79th Annual Meeting of the Transportation Research Board, Washington, DC
7. Farradyne PB (2000) Traffic incident management handbook. Federal Highway Administration, Office of Travel Management, Washington, DC
8. Region 4 ATMS local evaluation report (2005) Dunn Engineering Associates
9. Benefits of traffic incident management (1996) National Traffic Incident Management Coalition, Washington, DC
10. Project development manual, Appendix 6. New York State Department of Transportation. https://www.nysdot.gov/portal/page/portal/divisions/engineering/design/dqab/pdm
11. Klein LA (2001) Sensor technologies and data requirements for ITS. Artech House, Boston, MA
12. May AD (1990) Traffic flow fundamentals. Prentice-Hall, Englewood Cliffs, NJ
13. Gordon RL (2007) Design ITS user's manual

14. Martin PT et al (2001) Incident detection algorithm evaluation. University of Utah, Utah
15. Automatic incident detection algorithms (2001) ITS Decision Report. Partners for Advanced Transportation and Highways (PATH), University of California at Berkeley, Berkeley, CA
16. Mouskos KC et al (1999) Transportation operations and coordinating committee system for managing incidents and traffic: Evaluation of the incident detection System. Transportation Research Record 1679:50–57
17. Zhao H (2004) New York State Department of Transportation Inform system incident and communications log application user's guide. Dunn Engineering Associates
18. ARTIMIS operations plan (1997) The Advanced Regional Traffic Interactive Management & Information System, Cincinnati, OH
19. The NTCIP guide updated version 3 (2002) American Association of State Highway and Transportation Officials, Institute of Transportation Engineers, National Electrical Manufacturers Association
20. Ogden ME (2004) Guide to the IEEE family of standards. The Institute of Electrical and Electronics Engineers, New Jersey
21. Integrated Incident Management Systems (IIMS) (2005) New York State Department of Transportation, Region 11, http://www.dot.state.ny.us/reg/r11/iims/index.html. Accessed 24 Mar 2005
22. Gordon RL (2007) Incident management. Thinking Highways 2(3):42–46
23. Atkins C, Coleman M (1997) The influence of traffic calming on emergency response times. ITE J 67(8):42–46
24. Guidelines on traffic calming devices Street and Traffic Division, Public Works Department (2006) City of Kansas City, MO
25. Traffic incident management (TIM) self assessment national executive summary report (2003) Federal Highway Administration, Washington, DC

Chapter 5
Nonrecurrent Congestion: Incident Information to Motorists

Abstract Chapter 4 discussed delay reduction resulting from the decrease in response time to incidents. This chapter describes the other significant ITS approach to mitigating nonrecurrent congestion, namely providing motorists with information. Since messaging policies and diversion effects for recurrent and non-recurrent congestion may be different, these subjects are covered in this and the following chapter. For example, many agencies do not provide messages for recurrent congestion (see Chap. 6).

Motorist information can mitigate non-recurrent congestion in the following ways:

- Induce a motorist traveling on the freeway to divert in the event of an incident. Circumstances that induce diversion include:
- Capacity reducing incidents on freeways on which the motorist is traveling or on freeways that are accessed by the freeway upon which the motorist is traveling.
- Capacity reducing incidents, construction, or special events on remote facilities. Avoiding the incident on the remote facility may require the motorist to modify his route on the facility that he is currently using.
- Prior to the start of a trip or early in the trip, induce a motorist to change his travel mode or the start time of his trip.
- Make the motorist aware of unusual conditions such as incidents, lane blockages, and lane closures. Accidents may be reduced by facilitating the motorists' earlier preparation for these events.

Section 3.1.4 discusses the general issues relating to diversion resulting from non-recurrent congestion, and shows several diversion curves. It also discusses policy issues related to motorist information for nonrecurrent congestion.
The following topics are covered in connection with diversion to avoid incidents:

- The techniques for communicating with motorists.
- Strategies and policies for developing messages and the use of the content and the strength of the message to influence the percentage of motorists diverting.
- Semiautomatic and manually implemented strategies for controlling diversion.

R. Gordon, *Intelligent Freeway Transportation Systems: Functional Design*,
DOI 10.1007/978-1-4419-0733-2_5,

- The effect of diverted traffic on alternate routes.
- Models for delay reduction resulting from diversion for diverted traffic, nondiverted traffic remaining on the freeway and corridor traffic.
- The effect of diverted traffic on alternate routes and the necessity for controlling the impact.

The chapter provides guidance and some simple models to assist the engineer in locating changeable message signs. The importance of the quality of motorist information is discussed, and a simple evaluation measure is provided. The use of ITS for emergency evacuation is introduced and a model for generating and disseminating information is provided.

5.1 Motorist Diversion

5.1.1 Motorist Messaging Techniques

5.1.1.1 Technologies for Communicating with Motorists

Traffic information is often developed by agencies operating highway facilities. In some cases, the agency contracts with a private service to provide this information. Private services may also supply this information directly to motorists. Table 5.1 identifies a number of technologies that provide communication with motorists.

Information dissemination techniques may be provided by the operating agency directly to the motorist through changeable message signs (CMS) and by using highway advisory radio (HAR) transmitters in combination with vehicle radios. Other operating agency initiatives include 511 telephone information service that may be accessed in the vehicle or prior to the trip. These services, provided by many state departments of transportation, offer detailed traffic and roadway condition information along with other types of information.

Privately provided traffic information includes commercial radio station traffic broadcasts, satellite radio, television traffic condition reports, and in-vehicle GPS based navigation systems that provide real-time routing information. The traffic information updates for these navigation systems may be provided by cellular telephone or by satellite radio. The IntelliGuide program is an emerging technology for providing in-vehicle information using both roadside equipment and information from other vehicles.

Communication of traffic information may result in the motorist changing his lane use, route, mode or the time he initiates his trip. Most mode and trip initiation

Table 5.1 Techniques for providing motorist information

Technique	Message coverage characteristics	Limitations	Speed of response to incidents	Cost to motorist
Changeable message sign (CMS)	Coverage of roadway and possibly nearby roadways as generally used	Message length limited by readability considerations	Good	Nothing
Highway advisory radio (HAR)	Coverage radius typically 3–5 miles. In some cases may be used for pretrip planning.	Message length limited by transmission coverage	Good	Nothing
Commercial radio (conventional radio and satellite radio)	Wide area. May be used for pretrip planning	Completeness of conventional radio incident coverage often limited by broadcast time allocation. (Satellite radio generally provides better coverage)	Fair	Nothing
GPS in-vehicle navigation system (real-time information provided by satellite radio or cellular telephone based services)	Wide area	No effective message length limitation	Varies depending on traffic information source	Initial equipment cost plus service cost
E 511 telephone service	Wide area. Many states provide information organized by roadway section or other means to obtain specific information. May be used for pretrip planning	Selection of desired information may cause motorist distraction.	Varies depending on traffic information source	Cellular telephone service cost
IntelliDrive[a]	Locations where infrastructure equipment is deployed		Good	In-vehicle equipment plus any service charges
Web site	May include 511 and other services	Primarily useful for pretrip planning	Varies depending on traffic information source	Nothing

[a]At this writing IntelliDrive is in the planning stages. Deployment plans, costing and institutional arrangements have yet to be finalized

time changes occur prior to starting the trip. Chapters 5 and 6 primarily discuss route diversion issues that occur when the trip is already in progress.

5.1.1.2 Diversion Messages

Dudek indicates that operating agencies should only provide messages that contain information that the motorist does not already know, however, travel time messages are recommended by FHWA and are becoming increasingly popular [1]. According to this policy, messages should only deal with nonrecurring events. Many operating agencies implement this policy.

Messages that describe abnormal traffic conditions often result in the motorist diverting to another route. Diversion messages may be of either of the following types:

- Explicit diversion – Sometimes called active diversion, these messages indicate the need to divert, and may suggest an alternate route. In Table 3.1, the messages related to message strengths 4, 5, 7, and 8 are active diversion messages, and generally induce larger diversion levels than do implicit (or passive) messages. Dudek [1] suggests that the positive guidance policy described in Sect. 3.1.4.2 be implemented in the case of explicit diversion.
- Implicit diversion – Messages that describe such events on the roadway as accidents, lane blockages and closures, construction, special events, delay, and travel time above expectations may, depending on the delay that the motorist perceives, result in diversion. They are implicit or passive diversion messages. Messages 1, 2, 3, and 6 in Table 3.1 are implicit diversion messages.

5.1.2 Operational Diversion Policies and Strategies

When an incident occurs on a freeway, diversion options include:

- Diversion around the incident and return to the freeway downstream of the incident.
- Diversion to alternate local routes when these more conveniently access the motorist's destination.

While in some cases direct diversion to another freeway may be possible, most diversion opportunities require use of the surface streets for some, or all, of the diversion route. Surface street diversion routes often have less spare capacity than freeway diversion routes. During peak periods, both types of facilities may have little spare capacity.

During nonpeak periods, even modest diversion levels have the potential to congest surface street arterials that may serve as alternates. A simple example

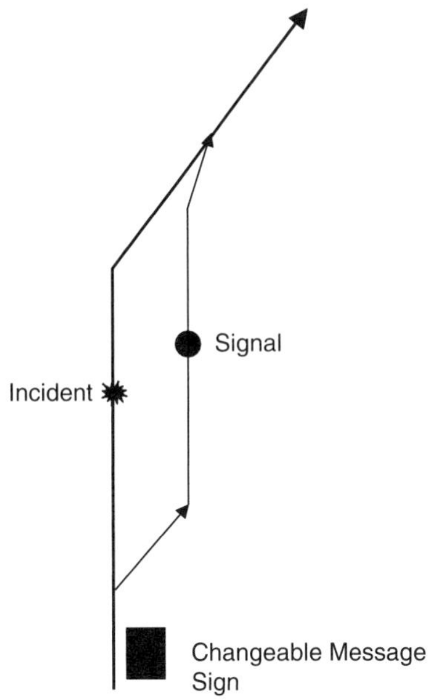

Fig. 5.1 Flow diversion during a nonpeak period

Table 5.2 Parameters for nonpeak period diversion example

Parameter	Symbol or relationship	Value
Three lane freeway (one direction)		
Normal capacity	C	6,300 VPH
Number of lanes blocked		1
Residual capacity	RE=0.49•C	3,087 VPH
Normal freeway volume	q	4,000 VPH
Capacity deficit	CD=q−RE	913 VPH
Arterial alternate (2 moving lanes)		
One direction roadway capacity	RC	3,600 VPH
Green signal split along arterial	GS	0.55
Capacity along arterial with signal	CS=GS•RC	1,980 VPH
Maximum flow without major arterial congestion	MF=0.9•CS	1,782 VPH
Normal arterial volume	qA	1,370 VPH
Maximum divertible freeway flow without major arterial congestion	MDF=MF−qA	412 VPH
Maximum diversion fraction for no major arterial congestion	DF=MDF/q	0.103

illustrating this statement is shown by the midday scenario in Fig. 5.1 and in Table 5.2.

Figure 5.1 depicts a single lane freeway blockage during a midday period. A medium capacity arterial alternate is directly accessible from the freeway and returns to the freeway downstream of the incident. There is a CMS upstream of the incident that has the capability to divert traffic. The residual capacity of the freeway under incident conditions is 49% of the basic capacity (see Table 4.1). This results in a capacity deficit of 913 vph.

With the signal, the alternate roadway has a capacity of 1,980 vph. Section 3.1.4.2 indicates that significant congestion on the surface street alternate is likely when the volume to capacity ratio exceeds 0.9. If a *policy* is adopted that does not permit excessive congestion on the alternate during nonpeak periods, then only 1,782 total vph may be accommodated, and only 412 vph may be diverted from the freeway.

Such a policy may be needed to

- Maintain a lower travel time on the alternate route than on the route with an incident.
- Preserve acceptable conditions for traffic that normally uses the alternate route.

The *strategy* to implement this policy requires the diversion messages to limit the diversion volume to 412 vph or a diversion fraction of approximately 10%. Section 3.1.4.1 indicates that the propensity to divert can be controlled through the "strength" of the message. The low diversion fraction required for this example implies that only a low strength message (Table 3.1) would be appropriate. The example illustrates the need to closely monitor traffic conditions on the alternate route to assure that the strategy is being properly implemented.

5.1.3 Strategic Network Management

Possible alternative policies include the following:

- Accept considerable congestion on alternate. The strategy to implement the policy in the example of Sect. 5.1.2 could result in raising the acceptable volume to capacity ratio to 0.96, resulting in a diversion volume of 531 vph, corresponding to a diversion fraction of approximately 13%. Transit on the arterial is also likely to experience considerable delay.
- Favor reducing delay on the freeway to the extent possible. Further increase in the diversion fraction will considerably increase delay on the arterial.
- Implement a strategy that minimizes traveler delay in the corridor.

By providing messages to motorists, agencies employ one or more of these policies. In some cases the agency has made a decision to adopt a policy while in other cases the selection of message formats implicitly defines a policy.

Policies may be supported by the traffic assignment and distribution models that are traditionally used for transportation planning. These models have a demand component (trip generation and traveler behavior) and a supply component (network characteristics). A model to optimize flow under changing network conditions requires more rapid adaptation of both the demand and supply component to short term changes. Dynamic traffic assignment (DTA) models address this need. For example, DynaMIT [2] models short-term changes in traveler demand and network characteristics. In addition, it accepts real-time inputs from field surveillance data and from traffic controls.

The diversion fraction (DF) is a key variable in diversion strategies (Sect. 5.1.4) that implement these policies. DF is determined by many of the issues previously discussed. From an operational standpoint it is best determined by measurements of traffic flow in response to messages. While DF is difficult to model for the evaluation of policies and strategies, for concept design purposes the Design ITS program [3] approximates DF as

$$\mathrm{DF} = \mathrm{H1} \bullet \mathrm{H2} \bullet \mathrm{H4} \bullet \mathrm{P21} \bullet \mathrm{P3} \bullet \mathrm{P4} \bullet \tag{5.1}$$

where

H1 = Probability of incident detection (see Sect. 4.6.3.1)
H2 = Probability of incident verification (see Sect. 4.6.3.1)
H4 = Probability of tail of queue and incident termination detection (see Sect. 4.6.3)
P21 = Probability that the TMC is manned when incidents occur
P3 = Probability the motorist receives and understands the message (*media factor*)
P4 = Probability that the qualifying motorist receiving and understanding the message diverts.

P3, the *media factor*, is discussed in Sect. 5.1.3.2. P4 consists of two elements, the *presentation factor* and the *content factor*. These are discussed in Sect. 5.1.3.1.

5.1.3.1 Control of Diversion Fraction Using Message Strength and Content

Section 3.1.4 indicates that the strength of the message *content* leads to different diversion probabilities. Peeta et al [4] suggests that message strength may be used to control the level of diversion, and provides the following binomial logit model as a basis for this control.

$$\mathrm{PD} = 1/\left(1 + e^{-U}\right) \tag{5.2}$$

where

PD is the propensity to divert

U is the difference in utility between diverting and not diverting motorists

In Peeta's formulation

$$\mathrm{U} = \mathrm{KD} + \mathrm{NMV} + \mathrm{MV} \tag{5.3}$$

where:

KD = a constant

Table 5.3 Propensity to divert based on message strength

Parameter	Message characteristics	Parameter value	Presentation factor (provided by author)
KD		−1.88	
NMV		+0.54	
MV for different messages shown in the following rows		Message type parameter (MV)	
Message 1	Occurrence of accident	0	0.20
Message 2	Location of the accident	−0.09	0.19
Message 3	Expected delay	+0.61	0.32
Message 4	The best detour strategy	+0.82	0.37
Message 5	Location of the accident and the best detour strategy	+2.08	0.67
Message 6	Location of the accident and the expected delay	+2.49	0.75
Message 7	Expected delay and the best detour strategy	+2.73	0.80
Message 8	Location of the accident, expected delay, and the best detour strategy	+3.55	0.89

NMV = a value that represents the utility for a set of variables that relate to factors other than the message type

MV = a value that represents the utility for the message type

Based on the message types of Table 3.1 and data provided by Peeta et al [4], Table 5.3 shows a set of model parameters for a particular set of motorist characteristics and the accompanying propensity to divert.

Because the utility parameters and the investigation methodologies vary among studies of this type, the data in Table 5.3 should be treated as an example of general trends.

The utility model underlying Table 5.3 is based only on the presentation of the message and does not include factors that relate to *content* (the delay time or the character of the incident) or the *media* (the technologies employed to deliver the message.) PD in (5.2) may therefore be considered as the *presentation factor*, and is only one factor in determining the diversion fraction. An example of the effect of the *content factor* is shown by the diversion curve in Fig. 3.3. P4 in (5.1) represents the combination of the content factor and the presentation factor.

5.1.3.2 Media Factor

The research to develop relationships such as Fig. 3.3 and Table 5.3 is generally based on stated preference surveys.[1] These surveys generally do not consider "real

[1] A stated preference survey is based on interviews, questionnaires, or in some cases on simulated conditions. A revealed preference survey is based on observation of actual motorist behavior.

world" conditions encountered by the motorist. In addition to the message content, the motorist's response to CMS messages also includes such issues as difficulty in understanding messages in locations unfamiliar to the motorist, language issues, or difficulty in observing the CMS due to occlusion by large vehicles and weather conditions, etc. All of these contribute to a reduction in diversion.

These issues reduce the probability that the motorist will divert in response to the message. Other information presentation media have similar issues. The probability that the motorist receives and understands the message from all available media is the *media factor* and is identified as P3 in (5.1).

Section 5.1.1.1 discusses various technologies for communicating with motorists. When more than one technology is available for message delivery, the media factor depends on the combination of the technologies that are employed. To illustrate how P3 may be estimated, the following model (based on [3]) considers the media factor using these message delivery techniques:

- Changeable message signs
- Highway advisory radio
- Other messaging choices such as those shown in Table 5.1 are treated below as a group for this illustration

P3=P31 when CMS are used without other messaging devices. P31 is given by

$$P31 = P34 \bullet P35 \tag{5.4}$$

P34 represents the probability that the motorist will encounter a CMS prior to reaching the congestion location as discussed in Sect. 5.2.2. P35 is the probability that the motorist will read and understand the message. Its default value in the Design ITS model [3] is 0.9.

P3=P32 when HAR is used without other messaging devices. The default value provided by Design ITS is 0.25.

P3=P33 when other message devices such as web sites, independent service providers, and traffic information provided by the media are used without CMS or HAR. The default value currently employed by Design ITS is 0.2. However, this value is expected to increase in the future due to the expanding use of 511 services, in vehicle real time traffic directions provided by GPS based navigators and the IntelliGuide initiative.

To combine the effects of the availability of multiple messaging media, the model uses the Bayesian inference technique similar to that used to compute (4.10). The relationships are as follows:

When CMS and other message devices are used

$$P3 = 1-(1-P31) \bullet (1-P33) \tag{5.5}$$

When HAR and other message devices are used

$$P3 = 1-(1-P32) \bullet (1-P33) \tag{5.6}$$

When CMS and HAR are used

$$P3 = 1-(1-P31) \cdot (1-P32) \tag{5.7}$$

When CMS, HAR, and other message devices are used

$$P3 = 1-(1-P31) \cdot (1-P32) \cdot (1-P33) \tag{5.8}$$

5.1.3.3 Operational Considerations in System Design

The preceding sections discussed the parameters that influence the diversion fraction through motorist information. System designs that consider these issues are more likely to result in more effective performance. The relationships leading to the diversion fraction are, however, complex, and the parameters that result in diversion are difficult to establish with certainty, and vary among roadway facilities.

From the perspective of the system operator, it is more appropriate to consider the diversion fraction in terms of a set of revealed preferences for each site for each set of motorist messages.[2] This is most effectively accomplished by measuring the diversion fraction with mainline and exit ramp point detectors, and statistically correlating the diversion fraction with the message set that results in the observations. While mainline point detectors are often included in ITS designs, exit ramp detectors are less often included. If the concept of operations includes diversion, it may be appropriate to consider inclusion of detectors necessary to measure the diversion fraction.

5.1.4 Diversion Strategies

Diversion policies may be implemented by the use of diversion strategies. Diversion strategies include open loop control (little or no capability for monitoring the controlled freeway and its alternate) and closed loop control. The following sections describe these strategies.

5.1.4.1 Open Loop Control

Strategy selection is based on policies that prescribe specific strategies corresponding to events such as incidents and their severity. Strategies may consist of diversion through motorist messages, and may be supported by traffic controls such as ramp metering, special signal timing plans, and lane use changes. Different prestored strategies may apply during different traffic periods. Simulation may assist in the development of these strategies. Open loop control is often used where little equipment

[2] A set of messages for each situation consists of a message for each medium employed for that situation.

for surveillance of traffic conditions is available. Open loop control provides little capability to adjust the control strategy in response to actual traffic conditions that develop during the operation.

5.1.4.2 Closed Loop Control

Closed loop control implies the adjustment of the diversion fraction and other traffic controls based on traffic observations by the system operator or by measurement by means of traffic detectors. In some cases, detector data are available for both the roadway with the incident and for the major alternate route (particularly when the alternate route is a freeway). In other cases, particularly when the alternate route is a surface street, appropriate information on the alternate is not available. The following sections describe several variations of closed loop control.

Closed Loop Manual Control

Figure 5.2 schematically shows this type of control. The operator, observing traffic conditions on displays of traffic detector, CCTV displays, and using incident information obtained from other sources, implements messages according to established policies.

Information from traffic detectors is often shown as a color-coded map display to simplify its interpretation by the operator. This is the most commonly used form of control when traffic surveillance equipment is available.

Surface street signal responses to changes in demand resulting from divertion may be improved by the use of adaptive traffic control systems. Selection of signal timing plans may be enhanced by recent advances in technology such as:

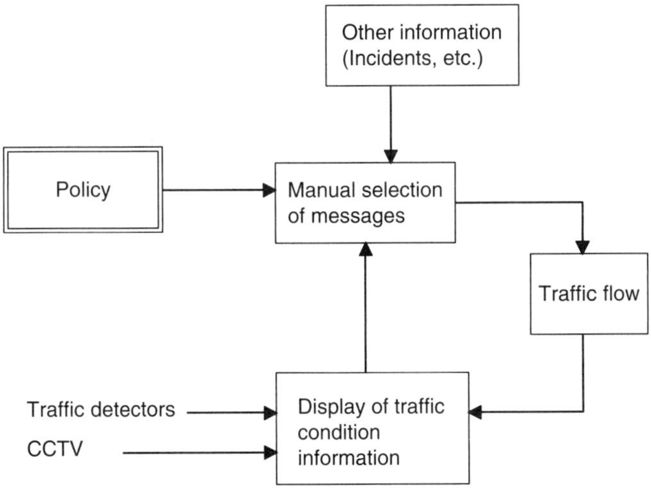

Fig. 5.2 Closed loop manual control

- Availability of delay information at traffic signals using conventional point detectors together with data processing techniques that coordinate detector data with traffic signal states to provide delay travel time and volume-to-capacity ratio.
- Probe based surveillance using toll tag readers, cellular telephone based vehicle location systems, or GPS based systems.

Closed Loop Semiautomatic Control

Examples of traffic condition states that may be used include speed, delay or traffic density. Implementing the control strategy automatically recommends a set of messages and controls based on policy rules and constraints. The message or controls may be modified by the operator based on additional information available to him such as CCTV observations and information from other sources. As with closed loop manual control, traffic condition information may be available on both the freeway being controlled as well as on the alternate route.

Figure 5.3 shows this type of control. Traffic detector information may be processed (possibly including the use of prediction) to include the traffic states that may be employed by the control strategy. Traffic condition information may be available on the freeway being controlled as well as on the alternate route.

Since closed loop semiautomatic control is implemented by traffic detector data, it is generally employed for both recurrent and nonrecurrent congestion.

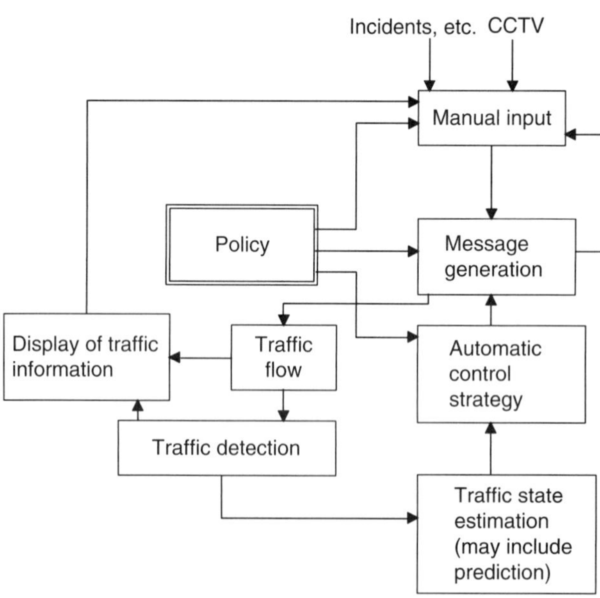

Fig. 5.3 Closed loop semiautomatic control

An example of closed loop semiautomatic control where the controlled freeway is monitored but the alternate route is not monitored is provided in Appendix D. An example with alternate route monitoring is described below.

Highway 401 in the Toronto, Ontario area features a set of express lanes and a set of collector lanes with transfer opportunities between these lane groups every few miles. The COMPASS system provides changeable message signs located on each roadway upstream of the transfer points. The CMS describe general traffic speed conditions on both routes. These speeds are measured by detector stations that are downstream of the transfer point. Figure 5.4 shows the general character of the control loop [5].

The diversion fraction between roadways is altered by providing messages on each CMS that indicate the conditions on both roadways. Table 5.4 shows the general COMPASS message structure for displaying these conditions on each roadway.

The COMPASS control strategy is a relatively simple example of closed loop semiautomatic control. This type of control may also be implemented with strategies that employ more detailed traffic models. For example, Kachroo and Osbay [6] provide mathematical models for dynamic traffic routing. They discuss the use of feedback-activated controls to optimize travel time and assure system stability.

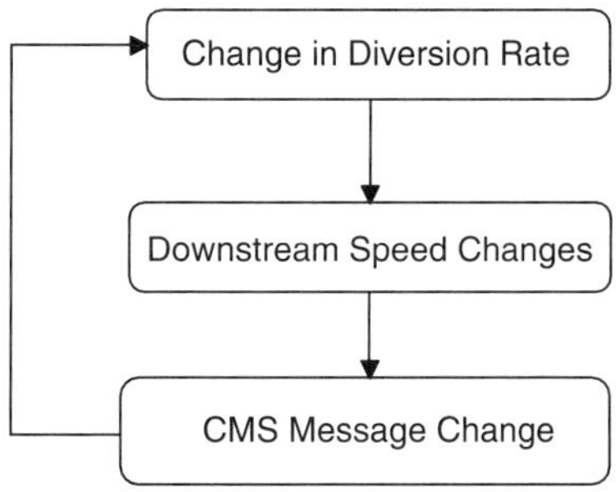

Fig. 5.4 Illustration of CMS feedback control loop (From Transportation Research Record: No. 2047, Transportation Research Board of the National Academies, Washington, DC, 2008, Fig. 5, p. 16. Reproduced with permission of the Transportation Research Board)

Table 5.4 Speed thresholds to determine message display where free flow speed is 100 km/h (62.1 miles/h) [5]

Current message	New message	Threshold speed
Moving well	Moving slowly	80 km/h (49.7 mph)
Moving well or moving slowly	Very slow	40 km/h (24.8 mph)
Very slow	Moving slowly	45 km/h (27.9 mph)
Very slow or moving slowly	Moving well	85 km/h (52.8 mph)

5.1.5 Reduction in Freeway Delay Resulting from Diversion

The most significant reduction in overall system delay resulting from diversion generally accrues to the freeway traffic that does not divert under incident conditions. Section 5.1.6 provides an example that illustrates this.

Figure 4.2 shows the effect of an incident on delay. Figure 5.5 is a similar figure that depicts the effect of diversion on the queues. It adds a line representing the new demand volume after diversion. The area enclosed by points a, b, c, and O represents delay under diversion conditions. Diversion commences after incident detection and confirmation (TC.) The vertical distance between the lines Oab and Ocb represents the number of vehicles in the queue at a given time.

The following definitions apply to Fig. 5.5 and to the following equations:

q_1 = Volume at incident clearance (roadway capacity)
q_2 = Volume entering incident location (prior to diversion)
q_3 = Volume when incident is present (capacity under incident conditions)
q_4 = Volume entering incident location after diversion
T = Time from incident occurrence until incident is cleared
TC = Time from the occurrence of the incident occurrence until diversion is implemented
TDD = Time period after incident clearance until queue clears under diversion conditions

Analysis of the geometrical relationships in Fig. 5.5 leads to the following equations

Delay reduced prior to TC

$$DD = 0.5 \bullet TC^2 \bullet (q_2 - q_3) \tag{5.9}$$

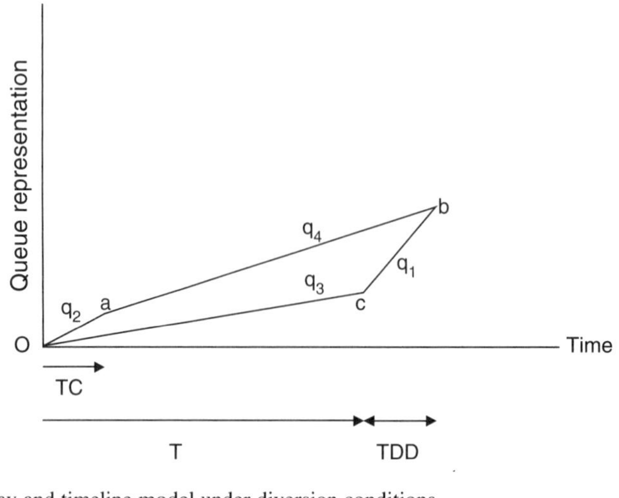

Fig. 5.5 Delay and timeline model under diversion conditions

Delay (DC) from TC to T

$$DC = 0.5 \cdot q_4 \cdot (T - TC)^2 + (T - TC) \cdot TC \cdot (q_2 - q_3) - 0.5 \cdot q_3 \cdot (T - TC)^2 \quad (5.10)$$

Time (TDD) from incident clearance to queue clearance

$$TDD = (q_4 \cdot (T - TC) + q_2 \cdot TC - q_3 \cdot T) / (q_1 - q_4) \quad (5.11))$$

Delay (DQC) after incident clearance

$$DQC = 0.5 \cdot TDD^2 \cdot (q_1 - q_4) \quad (5.12)$$

Total delay for nondiverted freeway traffic (DIF) is given by

$$DIF = DD + DC + DQC \quad (5.13)$$

Equation 4.4 describes the total delay, $D_{T,}$ with no diversion. The total time for the queue to clear is T + T_D where

T = Time from start of incident to incident clearance
T_D = Time for queue to dissipate after incident clearance (see Equation (4.1))

The total number of vehicles served during this period (NND) is given by

$$NND = q_2 \cdot (T + T_D) \quad (5.14)$$

and the average vehicle delay during this period (ADND) is

$$ADND = D_T / NND \quad (5.15)$$

Similarly, for the case where vehicles are diverted, the number of diverted vehicles served until queue clearance is given by

$$ND = q_2 \cdot (T + TDD - TC) \quad (5.16)$$

and the average vehicle delay during this period (ADD) is

$$ADD = DIF / ND \quad (5.17)$$

Figure 5.6 shows an example of a worksheet *Delay improvement on freeway* (included in the CD accompanying the book) to compute the key variables as well as DIF and the per-vehicle reduction in delay for nondiverting vehicles (IDV),

5.1.6 Effect of Diversion on Arterial Traffic

Preferred diversion policies and strategies require that:

- Motorists choosing to divert as a result of information explicitly or implicitly provided must benefit from such a diversion. Otherwise, the motorist may perceive that the information lacks credibility and may not use it in the future.

DELAY REDUCTION ON FREEWAY DUE TO QUEUE REDUCTION RESULTING FROM DIVERSION			
Required data entry			
		No Diversion	With Diversion
Capacity (Veh/hr)	q1 =	6000	6000
Demand Vol (Veh/hr)	q2=	4500	
Demand volume after diversion (veh/hr)	q4		4050
Residual Capacity After Incident (Veh/hr)	q3 =	3500	3500
Diversion Fraction	F	0.1	0.1
1-F	G	0.9	0.9
Blocked Time (hr)	T	1	1
Time to Detect & Confirm Incident (hr)	TC	0.07	0.07
Time after Incident Clearance Until Queue Clears (hr)	TD	0.67	
Delay Until Incident Clearance (veh hr)	D_T	500	
Delay Clearance to Queue Dissipation (veh hr)	D_Q	333	
Delay Until Detection (veh hr)	DD		2
Delay - Detection to Clearance (veh hr)	DC		303
Time after Incident Clearance Until Queue Clears (hr)	TDD		0.30
Delay after Clearance (veh hr)	DQC		87
Total Freeway Delay (veh hr)		**833**	**392**
Average Delay/Veh for Non-Diverting Vehicles (hr)		**0.111**	**0.074**
Improvement in Delay for Non-Diverting Vehicles (veh hr)	**DIF**		**441**
Improvement in Delay per Vehicle for Non-Diverting Vehicles (hr)	**IDV**		**0.037**
Improvement % for Non-Diverting Vehicles			**33.3**
Volume Diverted (veh/hr)	**VD**		**450**

Fig. 5.6 Worksheet computing delay reduced on freeway resulting from diversion

- The overall network must benefit from the diversion.
- The effects of diversion, particularly to surface streets, resulting from traveler information, must not exceed impacts that are agreed to in advance by the local stakeholders. Special signal timing plans may be used to facilitate traffic flow resulting from diversion and thereby reduce its impact.

A simplified example using a single diversion route illustrates these concepts. Using the example of Fig. 5.6, a freeway diversion fraction of 0.1 results in a diversion volume of 450 vehicles per hour.

The parameters for the example are provided in Table 5.5. The policy constraint in this example restricts diversion to 396 vph unless the signal timing is changed. In order to achieve the desired diversion level of 450 vph, the use of an alternative signal timing plan is required.

The time saved by each diverting vehicle is the time saved by nondiverting vehicles (IDV) plus the additional improvement in delay on the diversion route (TOND.) The total time saved by diverting vehicles (TSVD) is the product of this value and the number of diverting vehicles (ND.) The expressions for these relationships are provided as

$$\mathrm{ND} = \mathrm{VD} \bullet (\mathrm{T} + \mathrm{TDD} - \mathrm{TC}) \tag{5.18}$$

$$\mathrm{TSVD} = \mathrm{ND} \bullet (\mathrm{IDV} + \mathrm{TOND}) \tag{5.19}$$

where VD is the diversion volume.

Assuming a high strength message (Table 3.1), Fig. 3.3 shows that for the assumed diversion fraction of 10%, the delay saved by the diverting vehicles as compared with the nondiverting vehicles (TOND) is approximately 5 min. Using the example parameters yields

Table 5.5 Surface street diversion impacts for off-peak period

Parameter	Symbol or relationship	Normal green signal split	Diversion plan green signal split
Roadway capacity (VPH)	RC	3,600	3,600
Green split on arterial	G	0.4	0.5
Signal controlled capacity (VPH)	$\mathrm{CS} = G \bullet \mathrm{RC}$	1,440	1,800
Policy requirement	No significant off-peak additional congestion		
Maximum flow without significant congestion (VPH)	$\mathrm{MF} = 0.9 \bullet \mathrm{CS}$	1,296	1,620
Normal background traffic (VPH)	NFA	900	900
Maximum acceptable diversion flow without significant congestion	$\mathrm{MDF} = \mathrm{MF} - \mathrm{NFA}$	396	720

Green split shown excludes lost time (queue start up + clearance lost time)

ND = 572 diverted vehicles
TSVD = 68.6 vehicle hours of delay reduction for diverting vehicles

The delay to nondiverting vehicles is reduced by 441 vehicle hours (difference between no diversion and diversion delay in Fig. 5.6.) Since TSVD is considerably less than this value, the major system benefit for diversion is seen to result from benefits to nondiverting vehicles. For this example each diverting vehicle benefits considerably more than a nondiverting vehicle.

5.1.7 Reduction in Corridor Delay Resulting from Diversion for Incidents

Total corridor delay reduction (CDS) includes the sum of the delay reduced by nondiverting vehicles and diverting vehicles less the additional delay incurred by the prediversion traffic on the diversion route (DAR.) This is expressed as:

$$CDS = DIF + TSVD - DAR \tag{5.20}$$

This simplified model assumes that all traffic diverting from an incident will utilize the planned diversion route, and that prediversion traffic on the diversion route will also continue to use that route. A traffic assignment model provides a better estimate of the traffic actually using this route.

As diversion volume approaches the capacity of the alternate route, Fig. 3.4 shows that DAR begins to increase exponentially and the value of CDS may be significantly reduced.

5.2 Design Considerations for CMS Locations

Changeable message signs may take a number of forms ranging from simple blank-out signs and portable signs to large installations. Figure 5.7 shows some examples of the signs that inform motorists of non-recurrent traffic conditions such as construction, incidents, and weather conditions.

A CMS of this type requires a significant capital investment, and these CMS often represent a large portion of the cost of an ITS project. It is therefore important that the number and location of these CMS be selected based on cost effective principles. The exact location of a CMS depends on a number of factors such as the sight distance considerations, constraints on installation due to roadway construction features, nearby environmental issues, power access availability, and maintenance considerations. A number of these issues are discussed in Dudek [1]. The following sections treat the selection of CMS sites from a functional point of view, i.e. the effect of the CMS on notifying the motorist of incidents and other traffic related issues.

Fig. 5.7 (**a**) Changeable message sign displaying construction information (Source: Parsons Brinckerhoff, Inc.), (**b**) Small changeable message sign displaying incident information (Source: Daktronics, Inc.), (**c**) Changeable message sign displaying weather information and controlling speed (Source: Daktronics, Inc.)

5.2.1 Basic Considerations for CMS Functional Placement

The key functional objective for locating CMS is to place them so that the maximum number of viewers will be exposed to the CMS for incidents downstream of the CMS. An example of a methodology for implementing this principle on a network-wide basis is discussed by Abbas and McCoy [7]. Their model describes the optimal deployment of CMS for an entire network.

5.2.2 Simple Models to Assist in CMS Functional Placement

Many ITS projects are of limited scope, and address the requirements of a single freeway or a portion of that freeway. This section describes two simplified models that may assist in determining the functional placement requirements for CMS for that purpose. These models do not address the CMS placement requirements for providing messages for conditions on another roadway. The discussion adapts concepts from [7]. The general concept is shown in Fig. 5.8. The figure illustrates the simple case of a CMS located upstream of the last diversion opportunity prior to encountering the incident. The double-circled symbol in the illustration denotes the incident location and the checkered region represents the congestion building back from that point.

The value of the probability that the motorist encounters a CMS prior to the section containing the incident (P34) for Sect. A is given by the fraction of motorists who become candidates for diversion as a result of the CMS, prior to encountering the congestion and incident. In this case the value is

$$P34 = (VA - VA1) / VA \tag{5.21}$$

where

VA = Mainline volume upstream of queue formed by incident
VA1 = Entry volume downstream of CMS and upstream of incident.

Sections 5.2.2.1 and 5.2.2.2 discuss the simplified models for the following cases:

- Origin–destination data are available.
- Origin–destination data are not available

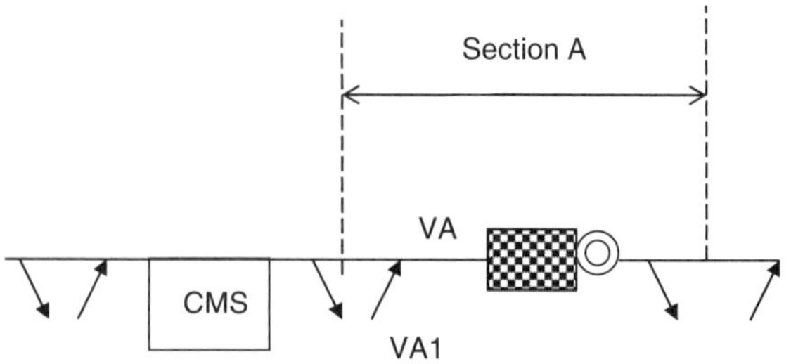

Fig. 5.8 CMS just upstream of traffic entering mainline before incident queue

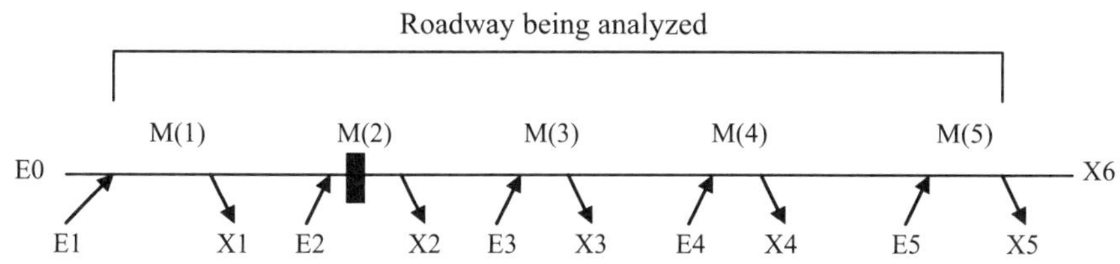

Fig. 5.9 Single roadway with one CMS

5.2.2.1 Computation of P34 When Origin–Destination Data Are Available

This section provides a model for computing P34 for a single freeway when origin–destination trip distribution information is available. This data may be obtained from the following sources:

- Planning models.
- Computer models that synthesize origin–destination data (trip tables) from traffic volumes.
- Special surveys employed for the collection of this data.

Consider the case where only one CMS is provided in the study roadway, and where no CMS is present upstream of the study roadway. This case is represented by the example shown in Fig. 5.9. The solid rectangle shows the CMS location.

Each of the E and X symbols represents a node at which the traffic enters or exits the roadway. Thus t_{23} represents the volume of traffic entering at node E2 and leaving at node X3. E0 and X6 are the mainline volumes entering and exiting the study roadway. A CMS is shown as a shaded rectangle in Sect 2. For an incident in Sect. J, diversion will take place for traffic observing the CMS and destined to exit at Sect. J and beyond. The computation of P34 for Sect. J for the case shown is of the form

$$\text{P34(J)} = \left(\begin{matrix} \text{Sum of the traffic volumes passing the CMS} \\ \text{and exiting at or downstream of Section J} \end{matrix} \right) / \text{M(D)} \tag{5.22}$$

where M(D) is the mainline volume for Sect. D which contains the CMS.

For the example shown

$$\text{P34(3)} = \left(\begin{matrix} t_{03} + t_{04} + t_{05} + t_{06} + t_{13} + t_{14} + t_{15} \\ + t_{16} + t_{23} + t_{24} + t_{25} + t_{26} \end{matrix} \right) / \text{M(2)} \tag{5.23}$$

$$\text{P34(4)} = \left(\begin{matrix} t_{04} + t_{05} + t_{06} + t_{14} + t_{15} + t_{16} + t_{24} \\ + t_{25} + t_{26} \end{matrix} \right) / \text{M(2)} \tag{5.24}$$

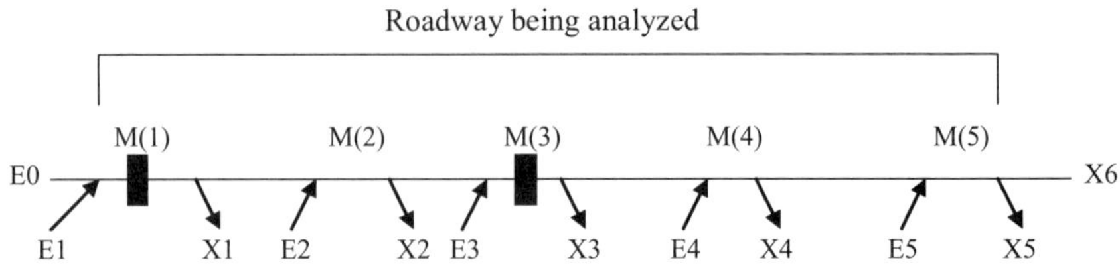

Fig. 5.10 Single roadway with multiple CMS

$$P34(5) = (t_{05} + t_{06} + t_{15} + t_{16} + t_{25} + t_{26}) / M(2) \tag{5.25}$$

When the roadways analyzed contain more than one CMS, each CMS provides diversion capability for incidents in the sections downstream of that CMS, until another CMS is encountered. This case is represented by Fig. 5.10, which shows CMS in mainline Sects. 1 and 3.

For the last CMS in the study area, diversion will occur for incidents in the sections downstream of that CMS. Qualifying traffic volumes are identified as follows:

- Origin node – All entering nodes between the upstream CMS and the subject CMS.
- Destination node – All exiting nodes starting with the section analyzed. Start with section downstream of the CMS.
- In terms of the example

$$P34(4) = (t_{24} + t_{25} + t_{26} + t_{34} + t_{35} + t_{36}) / M(3) \tag{5.26}$$

$$P34(5) = (t_{25} + t_{26} + t_{35} + t_{36}) / M(3) \tag{5.27}$$

For CMS upstream of the last CMS in the study area, qualifying traffic volumes are identified as follows:

- Origin node – All entering nodes upstream of the CMS. If there is a CMS upstream of this one, limit the origin nodes to nodes that are downstream of that CMS.
- Destination node – All exiting nodes starting with the section analyzed (downstream of the CMS) and extending to and including the section with the downstream CMS.

In terms of the example

$$P34(2) = (t_{12} + t_{13} + t_{14} + t_{15} + t_{02} + t_{03} + t_{04} + t_{05}) / M(1) \tag{5.28}$$

$$P34(3) = (t_{13} + t_{14} + t_{15} + t_{03} + t_{04} + t_{05}) / M(1) \tag{5.29}$$

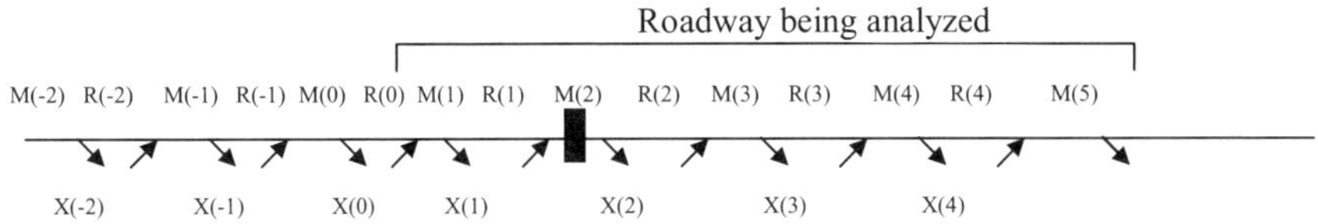

Fig. 5.11 Single Roadway with single CMS when no origin-destination data are available

5.2.2.2 Computation of P34 for a Single Roadway When Origin–Destination Data Are Not Available

This section describes the probability that the motorist encounters a CMS prior to a diversion point for an incident when origin–destination data are not available. It uses a simple trip assignment process based on freeway mainline and exit ramp volumes. The computation of P34 is based on a recursive process and is illustrated by the example of Fig. 5.11.

The diagram shows a CMS in Sect. 2. Definitions are as follows:

M(J) = Mainline AADT for Sect. J (volume between entry and exit locations.

X(J) = Sum of the exit ramp AADT for Sect. J.

R(J) = Residual volume after the exit ramps for Sect. J.

M_D = Mainline AADT for the section with the upstream CMS closest to the section being analyzed.

In addition to the sections in the roadway under analysis (section numbers higher than zero in Fig. 5.11), provision is made in the model for the possible location of the CMS in sections that are upstream of the roadway being analyzed. These are represented by sections numbered less than 1 in the figure.

For sections downstream of the CMS, the distribution model computes the traffic that has been exposed to the CMS after the section's exit point (denoted as P34(J)) as the product of the exposed traffic entering the section, R(J) and the ratio of the total remaining mainline traffic after the exit point to the total mainline traffic in the section (B_J). The equations that express this relationship are:

$$B_J = 1 - X(J)/M(J) \tag{5.30}$$

$$R(J) = R(J-1) \bullet B_J \tag{5.31}$$

$$P34(J) = R(J-1)/M_D \tag{5.32}$$

Figure 5.12 illustrates the computation of P34. This worksheet, *Computation of P34,* is provided on the CD accompanying the book. A "1" must be entered in the "CMS ON SECTION" column for each CMS on the roadway. Mainline volumes for a section containing a CMS upstream of the study roadway must be entered, along with volumes from that section to the study roadway. Exit ramp volumes for all sections (both upstream of the study region and in the study region (except for the last section) must be entered.

PROBABILITY THAT MOTORIST ENCOUNTERS CMS PRIOR TO INCIDENT (P34)

(shaded) Required entry deoending on roadway configuration

SECTION n	CMS ON SECTION	MAINLINE AADT	EXIT RAMP AADT	B	R	Basis	Intermed Comp	P34
-2	0	0	0	0.000	0	0		
-1	0	0	0	0.000	0	0		
0	0	0	0	0.000	0	0		
1	0	41800	4200	0.900	0	0	0.000	0.000
2	1	57200	5500	0.904	51700	57200	0.000	0.000
3	0	57150	5600	0.902	46634	57200	0.904	0.904
4	0	59750	5500	0.908	42341	57200	0.815	0.815
5	0	54700					0.740	0.740

Fig. 5.12 Worksheet showing computation of P34

Using these equations or the worksheet, it is possible to determine the most effective placement for CMS (based on the probability that the motorist passes a CMS prior to the section containing the incident) by analyzing various numbers and locations of CMS.

5.3 Quality of Motorist Information

Support for ITS ultimately rests, in large measure, on its customers, i.e. the motorists and their perceptions of the value of the service. Customer satisfaction is traditionally measured by stated preference surveys. Some states proactively solicit feedback on ITS as well as on other transportation services. Delaware DOT, for example, solicits information on the importance of a service. In 2006, the survey reported performance on "information on when to expect delay, road closings" as 4.79 on a scale of 1–7 [8].

In some cases, studies are conducted to identify the quality and value of particular services. Evaluation of 511 systems resulted in a customer satisfaction rate of 71% in Arizona, 68% in Washington State [9], 92% in the San Francisco Bay Area and 90% in Montana [10].

From the system designer's perspective, the quality or perceived benefit of motorist information depends on the following:

- Availability of technology and services to detect and confirm incidents and congestion related to incidents. Parameters H1, H2, and H4 in Sect. 4.6.3.1 represent these factors.
- Availability of traffic management center staff to implement messages at the time of the incident (a number of centers do not provide full time coverage). Parameter P21 in Sect. 4.6.3.3 represents this factor.
- The ability of the motorist to receive and understand the message. The media factor, P3 (Sect. 5.1.3.2) is a reasonable measure of this parameter.

Equation (5.33) shows how the Design ITS model combines these parameters to develop an approximate representation of relative quality (QUAL) of information for design purposes. QUAL is provided on a scale of 0–1.0.

$$\mathrm{QUAL} = \mathrm{H1} \bullet \mathrm{H2} \bullet \mathrm{H4} \bullet \mathrm{P3} \bullet \mathrm{P21} \tag{5.33}$$

5.4 ITS and Technology Applications in Emergency Evacuations

5.4.1 Introduction

Many agencies at all levels of government participate in emergency evacuation planning and operations. Successful culmination of these efforts involves interagency coordination and appropriate use of the assets available to the participating organizations.

Types of emergency evacuations include:

- Evacuations for which affected locations may or may not be anticipated in advance.
- The approximate time for the evacuation may or may not be anticipated in advance.

The types of situations that typically require evacuation include:

- Weather related incidents such as hurricanes.
- Hazardous material related incidents.
- Nuclear power plant incidents.
- Homeland security related events.

Emergency evacuations typically require planning and coordination by a number of agencies at all levels of government. The Freeway Management and Operations Handbook [11] provides an introduction to the subject.

5.4.2 *ITS and Technology Applications*

The principal role of ITS and ITS related technology to support emergency evacuations includes planning, traffic controls, and highway and traffic related information.

5.4.2.1 Planning

The following are common tools to assist in the planning of multiagency responses to emergency management and evacuation situations:

- By providing a common mapping reference for different agencies, the use of GIS systems facilitates the organization of emergency evacuations. Functions include disaster forecasts, vulnerability analysis, resource inventories, existing infrastructure inventory, shelter identification and status [12].
- Simulations to model traffic evacuation routes, demands, capacities, and emergency response traffic control measures. An example of a simulation available for this purpose is the Oak Ridge Evaluation Modeling System (OREMS) [13].

5.4.2.2 Traffic Controls

Elements of the evacuation plan may include traffic controls. Examples of traffic control applications include:

- Special timing provisions for traffic signals. Provisions may include signal timing plans that support evacuation routes, usually by means of longer green split periods along the evacuation route and longer cycle lengths. In some cases it may be appropriate to provide a constant green indication along the route.
- Contra-flow lanes on freeways. HOV lanes and reversible lanes may be used for this purpose, or conventional lanes may be reversed in support of an overall traffic management plan. In some cases ancillary traffic controls such as lane control signals and ramp access gates may be deployed to support the plan.
- Static signing is often used to identify emergency evacuation routes to motorists.
- CCTV is used to assist traffic management centers to support clearance of traffic incidents along emergency routes.

5.4.2.3 Highway and Traffic Related Information

The motorist information delivery technologies listed in Table 5.1 provide information during evacuations. During these events, the traffic management center serves as one element in the overall evacuation management process. The following discussion and Fig. 5.13 depicts an information dissemination model. It is abstracted from [14].

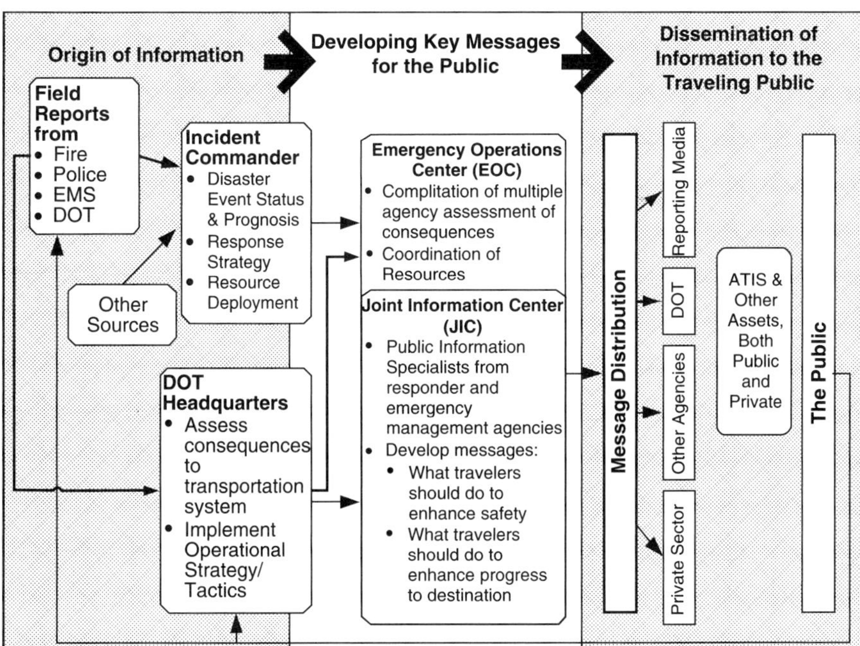

Fig. 5.13 Information dissemination model

The information flows depicted in Fig. 5.13 illustrate the full range of information generated and communicated to the public throughout a disaster situation. There are timing elements and dependency relationships that come into play as a disaster unfolds, as players enter or exit the picture and as information evolves. Moreover, the information flow embodies a continuous feedback loop, which incorporates the response of the public to the information received and the outcome of disaster mitigation.

References

1. Dudek CL (2006) Dynamic message sign message design and display manual. Report No. 0-40230P3 Texas Transportation Institute, Collage Station, TX
2. DTA System Enhancement and Evaluation of Traffic Management Center. (2000) Task P: framework for the use of DynaMIT-P. Massachusetts Institute of Technology, Cambridge, MA
3. Gordon RL (2007) Design ITS user's manual
4. Peeta S et al (2000) Content of variable message signs and on-line driver behavior. Transportation Research Record 1725: 102–108 Transportation Research Board, Washington, DC
5. Foo S et al (2008) Impacts of changed CMS messages on traffic diversion rates. Transport Research Record 2047 pp 11–18
6. Kachroo P, Ozbay K (1999) Feedback control theory for dynamic traffic assignment. Springer, London
7. Abbas MM, McCoy PT (1999) Optimizing variable message sign locations on freeways using genetic algorithm. 78th Annual Meeting of the Transportation Research Board, Washington, DC
8. Report on customer satisfaction, Delaware Department of Transportation, Dover, DE
9. Intelligent transportation systems for traveler information – Deployment benefits and lessons learned. U.S. Department of Transportation. http://www.its.dot.gov/jpodocs/repts_te/14319.htm. Accessed 27 May 2008
10. Americas travel information number –Implementation and operational guidelines for 511 services, version 3.0 (2005) 511 Deployment Coalition. Washington, DC
11. Neudorff LG et al (2003) Freeway management and operations handbook. Report FHWA-OP-04-003. Federal Highway Administration, Washington, DC
12. Pal A et al (2005) Enhancements to emergency evacuation procedures. UTCA Final Report 01105. University of Alabama, Tuscaloosa, AL
13. Oak Ridge evaluation modeling system (OREMS), Center For Transportation Analysis, Oak Ridge National Laboratory. http://cta.ornl.gov/cta/One_Pagers/OREMS.pdf. Accessed 27 May 2008
14. Communicating with the public using ATIS during disasters – Concept of operations (2006) Batelle and PBSJ, Columbus, OH

Chapter 6
Recurrent Congestion – Information to Motorists

Abstract Section 3.1.3 briefly described the differences between recurrent congestion and non-recurrent congestion. This chapter further elaborates on recurrent congestion. Discussion includes practices employed by traffic management centers and others in the provision of motorist information. The time periods when recurrent congestion is experienced, along with the variations in these periods, is discussed. The chapter covers diversion under recurrent congestion conditions, and describes the opportunities for diversion during shoulder periods.

6.1 Nature of Recurrent Congestion

Non-recurrent congestion, discussed in Chaps. 4 and 5, is generally caused by a temporary reduction in the normal capacity of the roadway. Recurrent congestion generally occurs when demand for facility use exceeds the facility's capacity and results in low average travel speeds, poor levels of service, and possible difficulty in access to and egress from the freeway. The demand trends and the resulting congestion are, within limits, generally repetitive. Severe recurrent congestion in major metropolitan areas is often experienced during peak commuting periods. Weekend shopping and recreation travel also often generate recurrent congestion. Congestion caused by special events is treated by some traffic engineers as recurrent congestion and by others as non-recurrent congestion.

6.2 Motorist Information During Recurrent Congestion

Section 5.1.1.1 describes the techniques and technologies commonly employed for providing motorist information. While the same technologies are used for providing information for recurrent and non-recurrent congestion, the types of information

R. Gordon, *Intelligent Freeway Transportation Systems: Functional Design*,
DOI 10.1007/978-1-4419-0733-2_6, © Springer Science+Business Media, LLC 2009

Table 6.1 Message content for recurrent and non-recurrent congestion

Technology	Recurrent congestion messages	Non-recurrent congestion messages
Information provided by state		
CMS	Delays, travel time, no delays, default messages, or blank sign representing no delays [a]	Location and nature of incident, delays, possible alternate route information, construction information
HAR	Rarely used	Location and nature of incident, delays, possible alternate route information, construction, weather, evacuation information
State website/511 website	Delays, travel time, traffic condition status map	Location and nature of incident, construction, weather, evacuation information
511 telephone	Delays, travel time [b]	Location and nature of incident, weather, evacuation information
Information provided by others		
Conventional and satellite commercial radio	Delays, limited travel time information. Reports often emphasize common congestion locations such as water crossings. Estimates of delays are sometimes provided	Location and nature of incident, construction information, weather, evacuation information
Real time GPS Information Service	Quickest route	Quickest route

[a] If policy requires provision of recurrent congestion information on CMS
[b] Some sites may indicate when conditions are clear

displayed may be somewhat different. Table 6.1 shows the types of information often provided by messaging technologies.

Dudek indicates that operating agencies should only provide CMS messages that contain information that the motorist does not already know [1]. Since most motorists have been repeatedly exposed to the same recurrent congestion in the same locations, many agencies choose not to display CMS messages describing recurrent congestion. Twelve percent of the agencies provide both types of messages (as well as messages that relate to safety and other items) [2]. Figure 6.1 shows a typical recurrent congestion message. When no congestion is present, the CMS may be left blank or a default message such as that shown in Fig. 6.2 may be used (motorists familiar with the system understand that such messages indicate that no congestion is present).

Semi-automatic message strategies of the type described in Sect. 5.1.4.2 are generally employed based on point or probe detectors. Messages for these strategies are generally developed without any consideration of the cause of the congestion

Fig. 6.1 Typical recurrent congestion message

Fig. 6.2 Typical default message implying lack of congestion

and thus they frequently provide recurrent congestion related information. When an incident occurs, in most cases the operator will replace the automatically developed message with another message that describes the incident, its location, and possible recommended remediation recommendations.

6.3 Variations During Periods of Recurrent Congestion

Recurrent congestion is common and pervasive during peak periods in major metropolitan areas, and is generally characterized by low speeds [3]. Variations in day-of-week demands, monthly demands, and random demands result in day-to-day speed and travel time variations for the same trip made at the same time of day.

As an example of these variations, the horizontal tics in Fig. 6.3 show the mean speed measured at a location in Portland, Oregon for weekdays from April 1, 2008 until April 11, 2008. The vertical bars represent the range of speeds encompassing one standard deviation above and one standard deviation below the mean. Thus, for the hour ending at 4 PM, the northbound detector station at milepost 304.4 reported a mean speed of 19.6 mph for this period and a standard deviation of 6.3 mph. This peak period also extends into the next hour. Although the mean speed is above 30 mph for the hour preceding and the hour following the peak period, the standard deviation is also high for these *shoulder* periods.

Standard deviation is a measure of travel time reliability. Provision of congestion, speed or travel time information enables the motorist to reduce his/her anxiety level by providing an indication, at least in some cases, of his anticipated delay. In some cases it provides diversion opportunities.

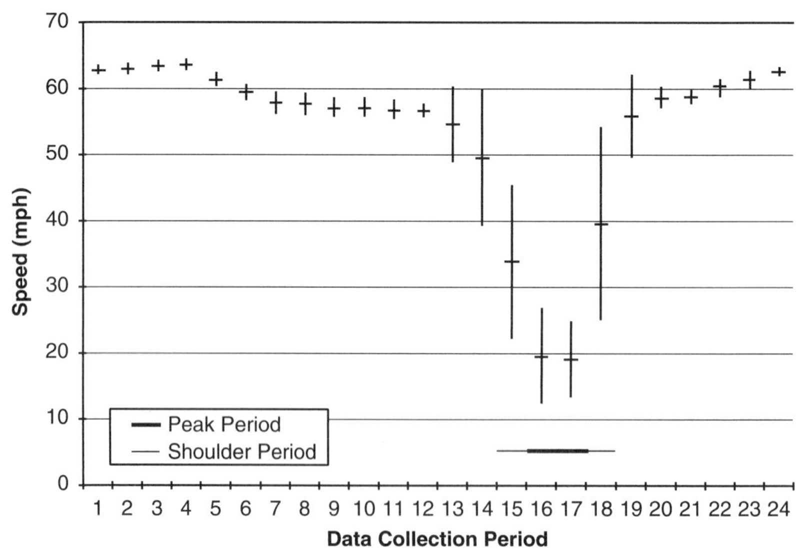

Fig. 6.3 Average Speed and Standard Deviation for I-5 NB milepost 304.4. Data for this figure were developed using the Portland Oregon Regional Transportation Archive Listing (PORTAL). PORTAL was developed by Portland State University under the direction of Dr. Robert Bertini.

6.4 Diversion During Recurrent Congestion

Congestion patterns in major urban areas generally follow Wardrop's principles (see Sect. 3.1.5). These principles indicate that although recurrent congestion delays may be significant, diversion in the absence of nonrecurrent congestion is not likely to be productive for either the individual motorist or for the entire system.

Wardrop's principles are, however, predicated on the assumption that conditions are invariant, and therefore that the motorist is aware of conditions on the routes in the travel corridor. Figure 6.3 and similar data, however, show significant levels of variation during these periods. This variation may present opportunities to reduce delay under some conditions.

Although speeds may be low during peak periods, speeds on the alternates are usually also low during these periods, so that significant diversion opportunities are relatively rare. During shoulder periods, however, there are times when the freeway speed is low, as shown in Fig. 6.3, and since volume-to-capacity ratios on the alternate routes are generally lower than during peak periods, there may be opportunities for diversion during these periods.

Because diversion opportunities are limited and delay savings resulting from CMS recurrent congestion messages generally amount to only a few minutes per trip, system-wide savings are usually modest. The savings are, however, generally achieved at little or no marginal cost.

References

1. Dudek CL (2004) Changeable message sign operation and messaging handbook. Report No. FHWA-OP-03-070, Federal Highway Administration, Washington, DC
2. Dudek CL (2008) Changeable message sign displays during non-incident, non-roadwork periods. NCHRP Synthesis 383. Transportation Research Board, Washington, DC
3. Shrank D, Lomax T (2007) The 2007 urban mobility report. Texas Transportation Institute, College Station, TX

Chapter 7
Ramp Metering

Abstract Ramp metering is often used in major metropolitan areas to reduce freeway delay and improve safety. It does this by (1) smoothing the flow of traffic at the merge and thereby increasing effective freeway capacity and (2) reducing the traffic entering the freeway which reduces the volume to capacity ratio. Topics covered by this chapter include:

- Ramp meter physical installation requirements and ranges of metering rates.
- Traffic flow breakdown models and the role of ramp metering in improving breakdown.
- Ramp metering strategies including pretimed, traffic-responsive, isolated and system-wide ramp metering.
- Ramp storage requirements and ramp queue control strategies.
- Ramp metering acceptance by the public.
- Ramp metering benefits model.

7.1 Introduction

Entry ramp control strategies include

- Ramp metering.
- Ramp closure.
- Special treatments including bus and carpool by-pass lanes on metered ramps.

These treatments and their implementation are described in the Freeway Management and Operations Handbook [1] and the Ramp Management and Control Handbook [2]. Ramp metering is the most commonly used of these treatments and is the subject of this chapter.

Ramp metering is implemented by a traffic signal on a freeway entry ramp (Fig. 7.1). By smoothing the flow at the merge with the mainline, it increases the service rate (bottleneck capacity) of the mainline, and reduces the accident rate. If the metering rate is established at a value that is below the average arrival rate at the ramp (restrictive ramp metering), a queue will build on the ramp causing additional

R. Gordon, *Intelligent Freeway Transportation Systems: Functional Design*,
DOI 10.1007/978-1-4419-0733-2_7,

Fig. 7.1 Ramp meter signal display

delay to the arriving vehicles. As a result, some of these vehicles will seek alternate routes, thereby reducing the demand volume at the entry ramp merge with the mainline. This, in turn, reduces the demand volume-to-capacity ratio at the merge and downstream of the merge and reduces the delay to the vehicles on the freeway mainline. Metering rates that are equal to the average vehicle arrival rate (non-restrictive ramp metering) build much smaller queues and generally do not result in significant volumes of traffic seeking alternate routes.

7.2 Background

7.2.1 *Early Metering Projects*

One of the first projects to establish the ability of metering to increase lane flow was conducted in the Lincoln Tunnel [3]. Lane changes are not permitted in the tunnel and bottleneck flows are frequently experienced at the foot of the upgrade in the tunnel. Early experiments showed that using fixed rate metering improved lane throughput from 1,200 cars per lane per hour to 1,320 cars per lane per hour. Once congestion sets in, the higher throughput could not be sustained. A later experiment

Table 7.1 Results of Lincoln Tunnel metering experiment

	Uncontrolled	Controlled
Average throughput (cars/lane/h)	1,210	1,290
Maximum throughput over a half-hour period	1,260	1,430
Average speed (feet/s)	27.2	40.9
Average density (cars/mile)	75.8	47.5

provided control based on volume and speed measurements. The results of the experiment are shown in Table 7.1.

Ramp metering projects started in the 1960s in Chicago, Detroit, Los Angeles, and Houston. These projects which included pretimed metering and traffic-responsive metering have been expanded over the years. Over 20 metropolitan areas in the United States currently operate ramp metering systems, and other parts of the world also use this technology.

7.2.2 Ramp Meter Installation Requirements

Several types of metering installations are employed depending on the metering rates and the configuration of vehicle storage capability on the ramp. For example, a single lane or multiple lanes may be metered. Metering may permit only one vehicle to pass the stop line per signal cycle or may permit multiple vehicles to pass (platoon metering).

Figure 7.2 [1] depicts a common deployment for a single lane, single vehicle meter. A standard 3-section (red-yellow-green), or 2-section (red-green) signal display is provided. The signals may be either mast arm or pole mounted. A sign or beacon is often used to indicate that metering is in effect.

For single vehicle meters, the metering rate is established by defining a metering cycle equal to the reciprocal of the desired metering rate. If the previous cycle has timed out (turning the signal to red), the signal will change to green when a vehicle is detected by the check-in (or demand) detector. When the vehicle is sensed by the check-out or passage detector the green interval is then terminated. The signal will remain in red until the traffic cycle times out at which time it will respond to the next arriving vehicle detected by the check-in detector.

Some ramp metering installations use merge detectors. The merge detector senses the presence of vehicles in the primary merging area of the ramp and freeway mainlines. When the merge detector senses a stopped vehicle blocking the merge area, the signal may be held in red for some preset maximum time in order not to congest the area and to reduce the possibility of a rear end collision [1].

One or more queue detectors are commonly used to prevent the queue from spilling back into the surface street traffic stream. Detection of vehicles by the

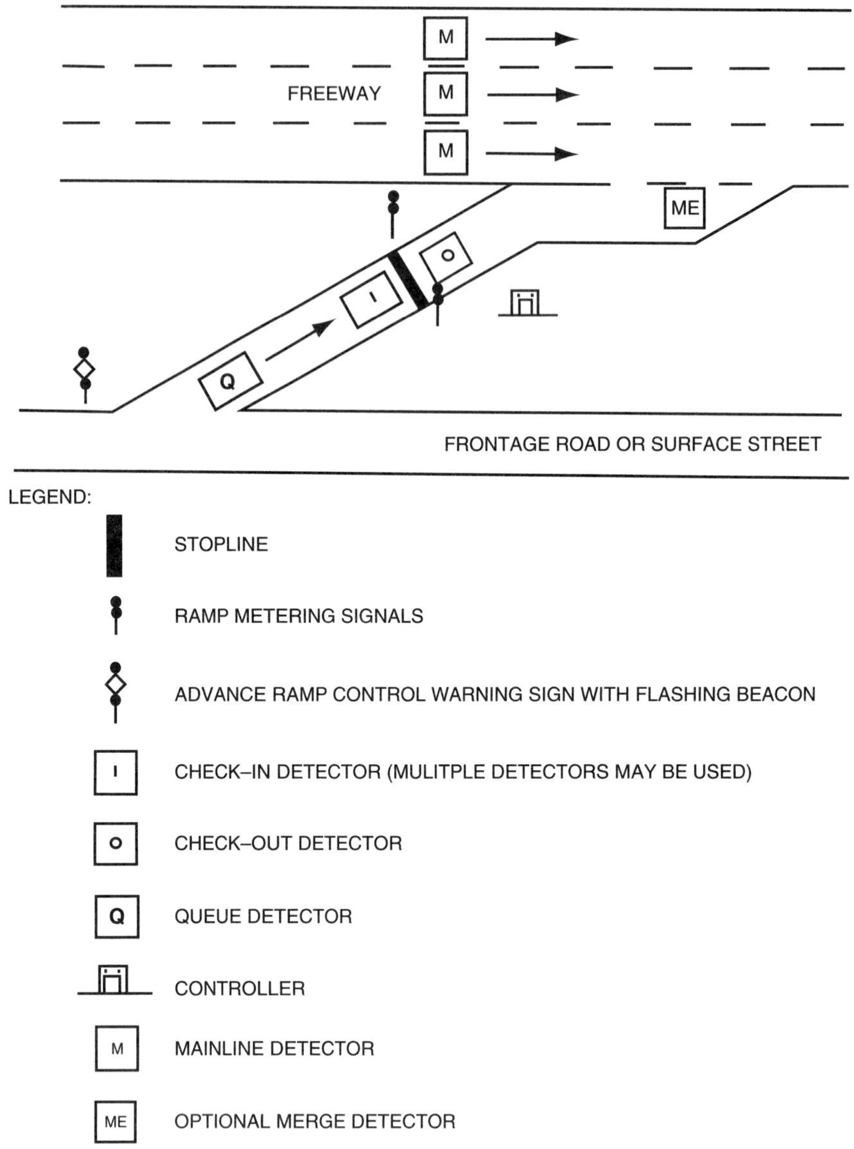

Fig. 7.2 Single-lane entry ramp metering system layout

queue detector increases the metering rate or terminates the metering. Strategies for accomplishing this are described in Sect. 7.4.5.2. In some cases the queue detector may be used to limit ramp waiting time to a specified value [1]. As discussed in Sect. 7.4.5.1, the lack of adequate vehicle storage capacity often limits the effectiveness of restrictive ramp metering.

Table 7.2 Ranges of ramp metering rates [13]

Types of metering	Number of metered lanes	Approximate range of metering rates (vph)	Comments
Single vehicle entry per green interval	1	240–900	• Full stop at the meter usually not achieved at 900 vph metering rate
Tandem metering – single vehicle entry per green interval per lane	2	400–1,700	• Applies when required metering rate exceeds 900 vph • Requires two lanes for vehicle storage • Vehicles may be released from each lane simultaneously or sequentially
Platoon metering – single lane multiple vehicle entry per green interval	1	240–1,100	• Platoon lengths permit passage of 1–3 vehicles per green interval • Principally used to increase metered volumes when geometrics do not permit use of more than one metered lane • Requires changeable sign indicating permitted number of vehicles in the green interval

Mainline detector placement is closely related to the particular control strategy implemented (see Sect. 7.4). To be viable for metering, the physical and traffic demand characteristics of ramps must lie within certain values. Deployment requirements for ramp metering are discussed in Sect. 7.4.5. Acceptable ramp meter rates are listed in Table 7.2.

7.3 Flow Characteristics and Freeway Capacity

7.3.1 Flow Characteristics for Near-Capacity Conditions

As traffic demand (i.e., volume) increases, density increases with a corresponding decrease in speed. As vehicle demand approaches highway capacity, traffic flow begins to deteriorate. Traffic flow is interrupted by periods of turbulence that disrupt efficiency. Traffic flow then begins to break down rapidly, followed by further deterioration of operational efficiency. An example of the breakdown in stable flow is shown in Fig. 7.3. The mean queue discharge flow is the volume departing the queue after flow breakdown.

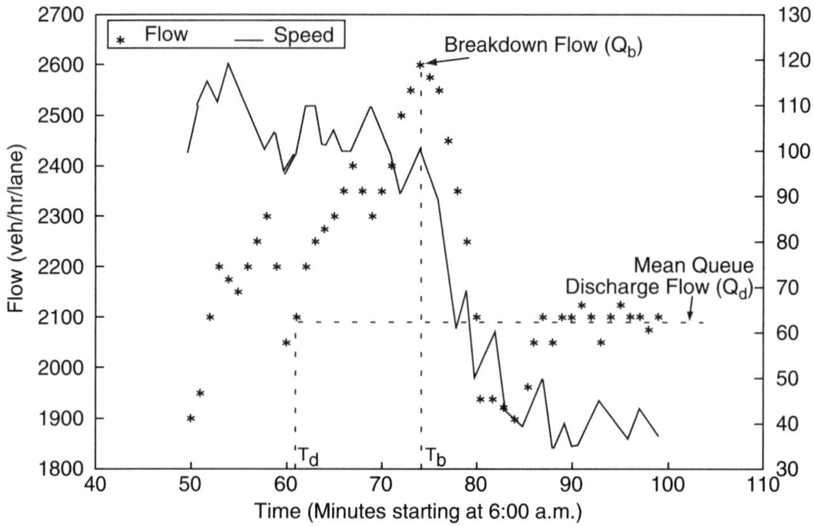

Fig. 7.3 Time trends for speed and flow during AM peak period. From, *Transportation Research Record: Journal of the Transportation Research Board, No. 1748*, Transportation Research Board of the National Academies, Washington, D.C., 2001, Figure 1, p. 111. Reproduced with the permission of the Transportation Research Board

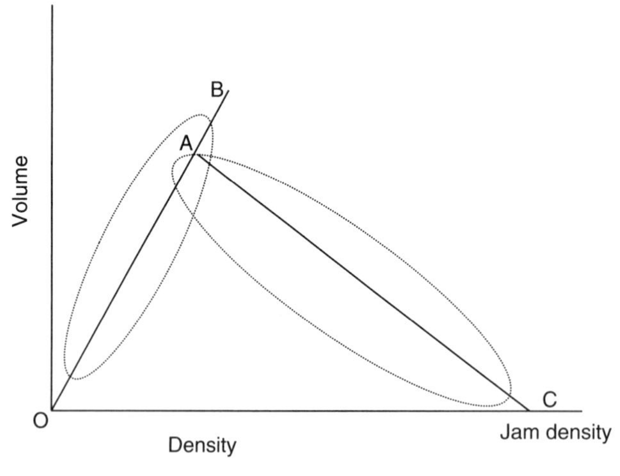

Fig. 7.4 Conceptual volume vs. density diagram

During the past decade there has been considerable research (e.g., References [4] and [5]) into identifying the properties of transitions between non-congested, stable flow and congested, unstable flow. Banks [5] suggests the models shown in Fig. 7.4. Line segments OA and OC constitute the inverted V model and line segment AB is added for the reversed lambda model. The dotted ovals indicate that considerable variation in the values of the actual data points is experienced. Jam density is the value of density when the traffic is stopped.

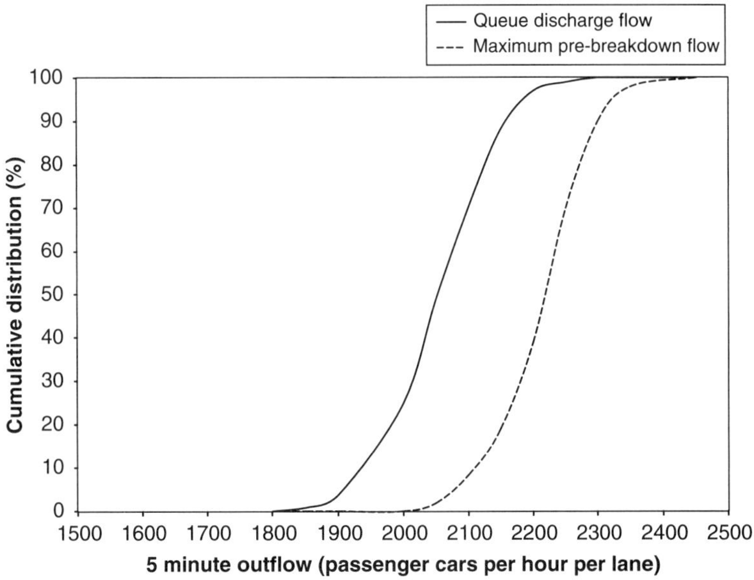

Fig. 7.5 Cumulative probability of distributions (redrawn). From *Transportations Research Record: Journal of the Transportation Research Board, No. 2012*, Transportation Research Board of the National Academies, Washington, D.C., 2007, Figure 6b, p. 14. Reproduced with permission of the Transportation Research Board

As volume increases, average density increases in an approximately linear relationship until the volume reaches the approximate location of Point A. This near linear relationship implies little speed change. When volume nears this point, the probability of the flow transitioning to an unstable state arises, and this generally characterized by lower volume, lower speed, and higher density. The location in a section of highway at which this transition first occurs is termed a bottleneck location. Bottlenecks typically occur at or near entry ramps as a result of merged traffic volumes that exceed roadway capacity. Line segment AC depicts the general trend in the unstable state, however, the actual data may vary widely from this trend. In Fig. 7.5 Shawky and Nakamura [6] show the best data point fit for the cumulative distribution for the maximum pre-breakdown flow and outflow rates (discharge flow from the queue after flow breakdown) on a section of a freeway in Tokyo.

Shawky and Nakamura [6] represent the cumulative probability distribution $P(x)$ by the Weibull logistic and normal functions of the form

$$P(x) = 1 - \exp(-(x/\beta)^{\alpha}) \tag{7.1}$$

where

x = outflow (queue discharge volume in passenger cars per hour per lane)

α = shape parameter

β = scale parameter

For the location used for Fig. 7.5.

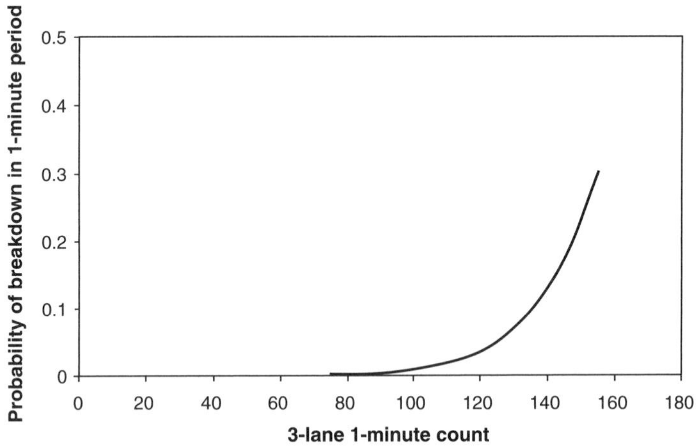

Fig. 7.6 Probability of flow breakdown for 1-min periods. From *Transportation Research Record: Journal of the Transportation Research Board, No, 1748*, Transportation Research Board of the National Academies, Washington, D.C., 2001, Figure 5, p. 113. Reproduced with permission of the Transportation Research Board

$\alpha = 28.7$
$\beta = 2{,}256$

Figure 7.6 illustrates an example of flow breakdown probability for the Toronto area for each one-minute period.

7.3.2 *Effective Capacity Improvement Through Ramp Metering*

This section describes the effect of ramp metering on the flow breakdown characteristics described in the previous section. The discussion applies the concepts and data described in Zhang and Levinson [7]. That reference examines the traffic flow characteristics at twenty-seven PM peak period active bottlenecks in the Twin Cities area with and without ramp metering. The motivation for this study is discussed in Section 7.5.2.

Figure 7.7 shows the model used by Zhang and Levinson [7] to analyze flow breakdown situations. As volume increases during the peak period and crosses the mainline queue discharge flow level, a prequeue transition period commences. The steady state queue discharge flow is shown by the dashed lines in the figure. The prequeue transition period lasts until the flow equals the steady state queue discharge flow. During this period, several instances of flow breakdown and restoration to pre-breakdown conditions may occur. In time, the flow decreases below the values shown in the figure as a result of the decrease in demand. Ramp metering delays flow breakdown and results in higher discharge rates after breakdown.

While the results vary considerably from ramp to ramp, Zhang and Levinson's results revealed the following average improvements:

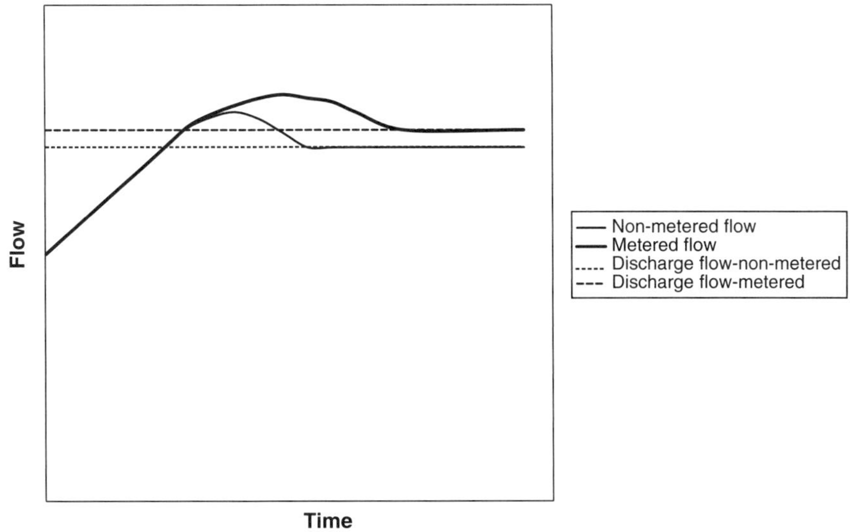

Fig. 7.7 Flow profiles at a bottleneck with and without ramp metering

- Ramp metering increased the average prequeue transition period from sixty minutes to nearly two hours, a 73% increase.
- The average flow during the steady state queue discharge period is 5.8% lower than during the pre-queue transition period. Thus the extension of the pre-queue transition period results in a considerable reduction in delay.
- The average value of the flow during the prequeue transition period is 3% higher with ramp metering.
- The average value of the flow during the steady state queue discharge period is 2% higher with metering.
- The average number of flow breakdown occurrences per afternoon peak period reduced from 1.2 without metering to 0.4 with metering. This occurs in part because of the elimination of flow breakdown at approximately half of the locations studied.
- Reducing the number of queues and their duration reduces the probability of spillback of these queues to the next upstream entry ramp.

7.3.3 Freeway Service Improvement Through Ramp Metering

Both non-restrictive and restrictive ramp metering improve freeway throughput, delay and safety in the following ways:

- Smoothing the merge flow through ramp metering reduces the accident rate. A survey of management centers in eight cities found that ramp metering reduced the accident rate by 24–50% [8]. The accident rate improvement includes the reduction of secondary accidents and the delays resulting from both primary and secondary accidents.

- A large number of studies report significantly reduced travel time resulting from ramp metering. For example, non-restrictive ramp metering on Long Island provided an improvement of 20% [8]. Restrictive ramp metering may provide additional benefits as described below.
- Ramp metering may provide throughput increases. For example, system-wide restrictive ramp metering resulted in a 10% increase in freeway volumes in the Minneapolis-St. Paul, Minnesota area [9].

Restrictive ramp metering provides an additional benefit resulting from the reassignment of traffic resulting in a decreased demand for the ramp. This concept is illustrated by the following simple example showing restrictive metering at a single ramp. Figure 7.8 shows a freeway (line FBD) and an alternate route (EACD). Prior to metering, the volume entering the ramp at location B exceeds the 5,000 vehicle per hour (vph) capacity of the mainline and causes a mainline queue to build as shown in the figure, resulting in a delay of 400 vehicle hours during the two hour peak period. Prior to metering, most of the vehicles on link EA that are destined for

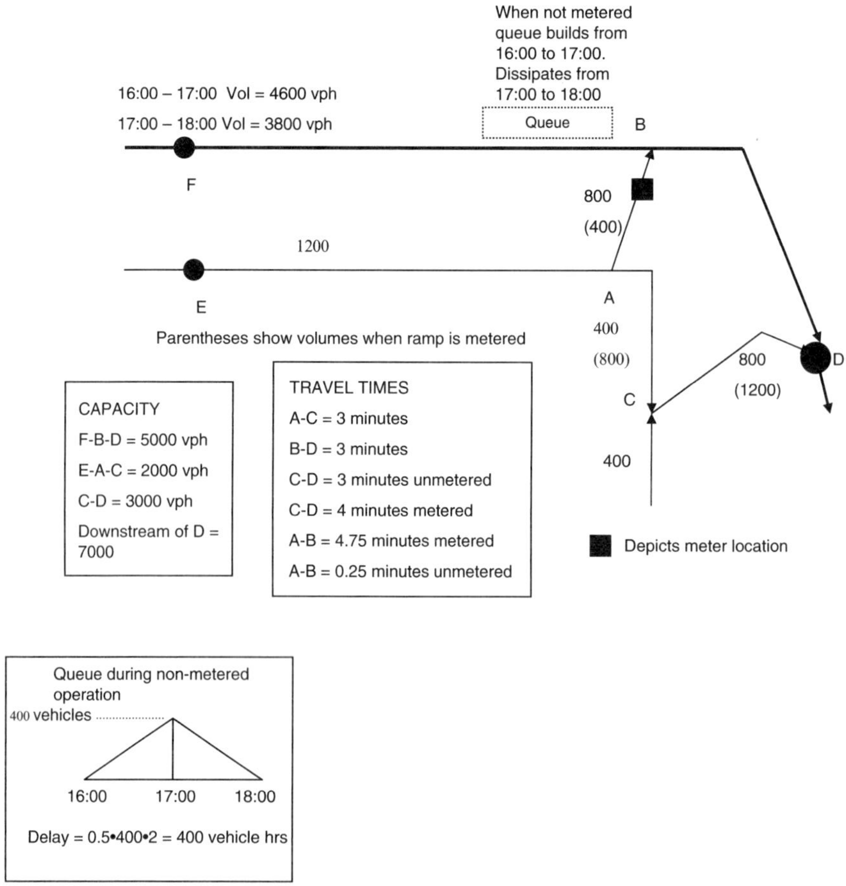

Fig. 7.8 Example of local restrictive ramp metering

location D and beyond choose to use the freeway because the trip is 2.75 min shorter. The downstream merge volume of reassigned traffic with the freeway traffic at location D results in a volume that is below the freeway capacity at this point.

Figure 7.9 contains the pertinent data for the metered and unmetered situations. With the initiation of ramp metering at an entry rate of 400 vehicles per hour, the queue builds on the entry ramp until the travel time via the freeway route and the alternate route are approximately equal.

The vehicle hours traveled for the segments (other than the mainline queue) were computed by the relationship:

VEHICLE HOURS FOR TRAFFIC WITH ROUTE CHOICES AT ENTRY RAMP A AND BACKROUND TRAFFIC ON ALTERNATE				
Unmetered entry ramp volume (UERV)			800	
Metered entry ramp volume (MERV)			400	
Base alternate traffic (BAT)			400	
Diverted alternate traffic (DAT)			400	
Traffic entering alternate at point C (CTA)			1000	
SEGMENT	Unmetered Alternative Trav time (min)	Unmetered Alternative Vehicle Hours	Metered Alternative Trav time (min)	Metered Alternative Vehicle Hours
AC	3.0	40.0	3.0	80.0
BD	3.0	80.0	3.0	40.0
CD	3.0	140.0	4.0	240.0
AB	0.3	6.7	4.8	63.3
Subtotal		266.7		423.3
Vehicle hours in mainline queue		400.0		0.0
Total - Travel time on alternate + mainline queue delay		666.7		423.3
System delay reduced by metering		243.3	vehicle hours	
Travel time for vehicles at point A with metering (minutes)				
By freeway	7.8			
By alternate	7.0			

Fig. 7.9 Worksheet for restrictive ramp metering example

Segment vehicle hours traveled = (segment volume) • (segment travels time in minutes/60) • (duration of peak period) (7.2)

Restrictive metering of this ramp results in a significant reduction in system delay. These benefits are typically not, however, uniformly distributed among all motorists. In this example, improvements resulting from the elimination of the mainline queue accrue to motorists on the mainline upstream of the ramp at location A. Motorists on the alternate route destined for location D and beyond, whether entering the freeway ramp, remaining on the alternate route, or entering the alternate route at a location downstream of the metered ramp experience a longer trip time.

7.4 Ramp Metering Strategies

The previous sections provide the general background on ramp metering and describe the mechanisms that ramp metering provides for relieving congestion. There are, however, potential negative effects that may result from ramp metering. These include:

- Additional delay to motorists who normally enter the ramp, even if they elect to use an alternate route.
- Additional delay to motorists who do not normally use the ramp but do use an alternate route.
- Possible spillback of traffic onto the surface street network.
- Motorist dissatisfaction resulting from the above.

The success of a ramp metering project depends, in part, on planning efforts to determine if metering is, in fact, feasible, and to select the metering strategy that best addresses the specific issues. Simulation is often a valuable tool for estimating the effects on the alternate routes, on the highway system in the vicinity of the metered ramp, and on the additional travel time that the ramp users and divertees will experience. It may assist in selecting the ramp metering strategy to be employed.

7.4.1 Overview of Metering Strategies

The metering rate for non-restrictive ramp meters is established at a higher level than the average arrival rate. The ramp queues are relatively short, consisting essentially of vehicle platoons released from an upstream signal or small vehicle platoons where the surface street upstream of the ramp is not signal controlled. The queue usually clears before the arrival of the platoon released by the subsequent signal cycle. The instrumentation requirements are relatively low (no mainline detectors are required) and communication with the traffic management center is not essential.

Restrictive ramp metering strategies include local metering strategies, system-wide metering strategies and pretimed and traffic-responsive metering strategies. Successful restrictive ramp metering treatments usually depend on detailed planning. Table 7.3 provides a summary of the major characteristics of restrictive ramp metering strategies.

7.4.2 Pretimed Restrictive Ramp Metering

Pretimed restrictive local ramp metering is appropriate for mitigation of recurrent congestion when the alternate routes downstream of the meter can accommodate the reassigned traffic and:

- Control of upstream ramps is not required to relieve congestion at the ramp merge or in the downstream section fed by the merge or
- It is not possible to meter upstream ramps.

A typical application is the mitigation of congestion at a single bottleneck as illustrated by the example in Sect. 7.3.3.

Pretimed system-wide ramp metering may be used to relieve recurrent congestion when:

- Metering a single ramp cannot provide a sufficient reduction in freeway volume. The lowest feasible metering rate at a ramp upstream of the bottleneck may not be sufficient to reduce demand to a level that is below bottleneck capacity. Metering of additional upstream ramps may be able to achieve or more closely approach this objective.
- The presence of multiple bottlenecks requires the consideration of metering at a number of ramps.

May [10] formulates relationships (termed demand–supply analysis) for a series of ramps in a manner that is generally similar to that in the example of Sect. 7.3.3. He describes the optimal control strategy for system-wide ramp metering as a linear programming procedure that maximizes mainline volume. His analysis treats the traffic diverted to the alternate route in a simple fashion.

Constraints in the linear programming formulation include the following:

- Volumes on each freeway link must be below capacity.
- Metering rates must fall within practical limits.
- It may not be possible to control some entry ramps. In this case the freeway entry volume equals the arrival volume at the ramp.
- There may be additional constraints such as limitations on the size of the ramp queue and limitations on the volume of re-assignable traffic resulting from capacity limitations on the alternates.

The design of this system-wide pretimed ramp metering strategy may be facilitated by the use of simulation [11].

Table 7.3 Characteristics of restrictive ramp metering strategies

	Pretimed		Traffic-responsive	
	Local metering	System-wide metering	Local metering	System-wide metering
Functions				
Safety improvements through flow smoothing	Useful	Useful	Useful	Useful
Capacity improvements through flow smoothing	Useful	Useful	Useful	Useful
Congestion at a bottleneck that can be mitigated by metering at a single ramp	Useful	Not required	Useful	Not required
Significant number of nonmeterable ramps between metered ramps	Useful	Can possibly improve local metering performance	Useful	Can possibly improve local metering performance
Congestion at a bottleneck requiring metering at a number of upstream ramps	Not useful	Useful	Not useful	Useful
Traffic reassignment on a long term (strategic) basis	Useful	Useful	Can possibly improve pretimed metering performance	Can possibly improve pretimed metering performance
Traffic reassignment on a short term basis (including nonrecurrent events)	Not useful	Not useful	Useful	Useful
Implementation Issues				
Manual data collection	Usually required	Usually required	Not usually required	Not usually required
Mainline detectors	Not required but may be used to establish data base	Not required but may be used to establish data base	Required	Required

Management by TMC and field communication with TMC	Useful but not required	Useful but not required	Required	Required
High level of vehicle storage capability	Required	Required	Required	Required
Ramp queue spillback protection	Required	Required	Required	Required
Capital and maintenance cost	Relatively low	Relatively low	Higher per metered ramp	Highest per metered ramp

7.4.3 *Local Traffic-Responsive Restrictive Ramp Metering*

Local traffic-responsive ramp metering provides the capability to adjust the metering rate to real time traffic conditions in the vicinity of the metered ramp. Local traffic-responsive metering is also employed as a component of a number of system wide traffic-responsive metering strategies.

7.4.3.1 Occupancy

By accommodating day-to-day variations and shorter-term variations in mainline traffic volume, local traffic-responsive ramp metering can improve performance relative to pretimed metering. Many ramp metering control algorithms use occupancy as the key parameter to establish metering rates.

Occupancy is the ratio of the time that vehicles occupy the detection zone of a traffic detector to a specified time period. It is sometimes used in ITS as a surrogate for traffic density (vehicles per lane per mile).

Different types of traffic detectors provide different vehicle sensing distances on the lane, that is, the occupancy time of a vehicle as sensed by a detector is

$$t_j = (LV+LD)/S_j \tag{7.3}$$

where

t_j = occupancy period sensed by detector for vehicle j

LV = length of vehicle

LD = length of detector sensing area

S_j = speed of vehicle j

The values of t are summed over a time period and divided by that time period to obtain the occupancy as reported by that type of detector according to:

$$\theta = (1/T) \bullet \sum_{j=1}^{m} t_j \tag{7.4}$$

where:

θ = occupancy over averaging period T as indicated by the detector

m = number of vehicles passing the detector during time period T

Shawky and Nakamura [6] relate flow breakdown probability to occupancy in the same way that they did for volume (Fig. 7.5). Figure 7.10 shows the best fit to the data points for a particular ramp merge. Shawky and Nakamura indicate that the relationship between flow breakdown probability and occupancy is more consistent among different ramp locations than is the relationship to volume, and therefore may be preferable to use as a control variable.

Figure 7.11 shows a plot of volume vs. occupancy data for one minute periods taken prior to the start of the PM peak period and continuing into that period. The trend lines in the figure may be compared to the trend lines in the volume vs. density plot of Fig. 7.4. The zero volume intercept of the congested condition trend line in Fig. 7.11 corresponds to the jam density point of Figure 7.4. Its value is less than

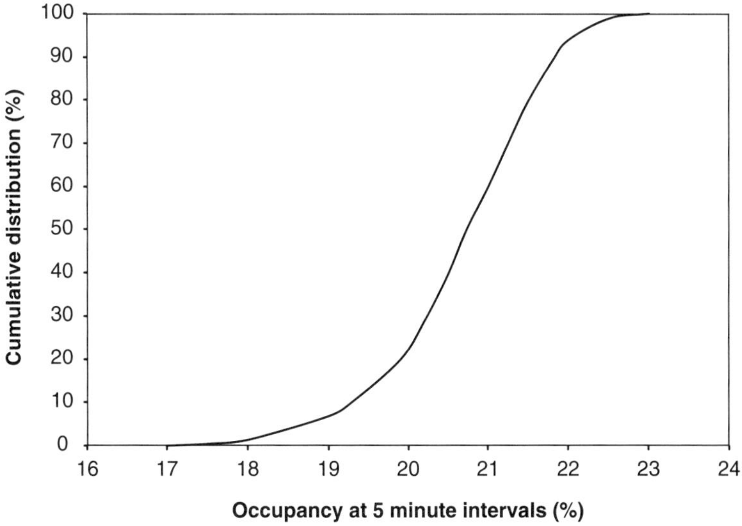

Fig. 7.10 Cumulative probability of flow breakdown vs. occupancy (redrawn). From *Transportation Research Record: Journal of the Transportation Research Board, No, 2012*, Transportation Research Board of the National Academies, Washington, D.C., 2007, Figure 8a, p. 16. Reproduced with permission of the Transportation Research Board

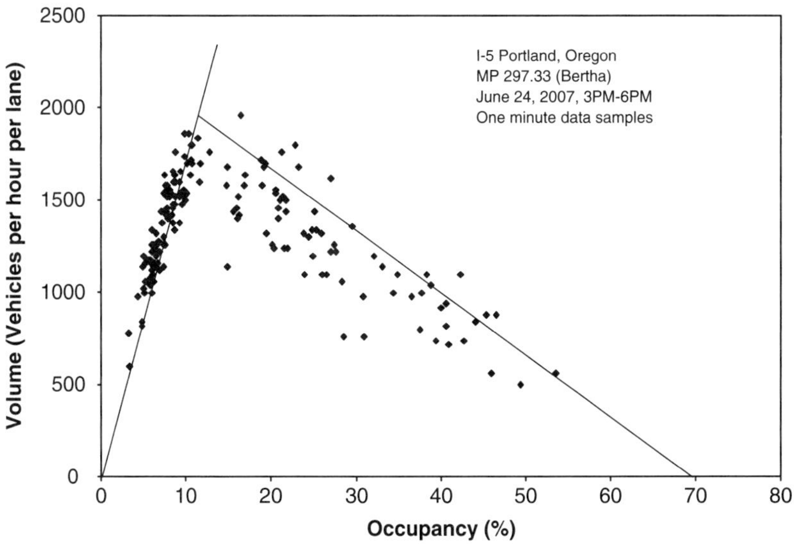

Fig. 7.11 Volume vs. occupancy data. (Developed from Portland Oregon Regional Transportation Archive Listing – PORTAL)

100% occupancy because, for most detectors, the apparent length that the detector itself contributes when each vehicle is sensed is less than the space between vehicles at jam density. For example, the sensed distance for the commonly used six foot square inductive loop detector is approximately six feet. The inter-vehicle distance at jam density is longer. Thus the average spacing between vehicles at jam density is greater than the detector's sensed distance along the roadway and an occupancy value of 100% will not be obtained even under highly congested conditions.

7.4.3.2 Explicit Scheduling of Metering Rate

One common control approach is to schedule the metering rate as a function of the occupancy measured during the previous time interval as shown in Table 7.4 [12]. Occupancy is measured by means of mainline detectors situated near the ramp merge with the mainline as shown in Fig. 7.2.

Occupancy data, however, exhibits considerable minute-to-minute variation, particularly when operation is in the transition region between the noncongested flow regime and congested flow. Figure 7.12 shows a portion of the data for one minute intervals at the same location as for Fig. 7.11. This data encompasses the transition period between noncongested and congested flow as well as a portion of the congested flow period.

If control were to be implemented using raw data, the current occupancy might be considerably different from the value during the previous interval, i.e., the interval whose data are used to set the parameters for the current interval's metering rate. This would result in considerable variation in the metering rate during successive intervals. In order to mitigate this problem, many ITS use occupancy data that are processed by a filtering or smoothing process. The relationship for a first order linear filter that is commonly used [13] is

$$\theta_O(j) = \theta_O(j-1) + K \bullet (\theta_I(j) - \theta_O(j-1)) \tag{7.5}$$

where:

$\theta_O(j)$ = filter output after the jth instant

$\theta_I(j)$ = filter input data value (average value of variable between $j-1$ and j instants)

K = filter coefficient in the range 0–1.0 (K = 1.0 represents no filtering)

Figure 7.13 shows that the large minute-to-minute variations in the unfiltered data plot are considerably reduced as filtering is increased (reduced value of K). However, the time lag in the filtered data increases as the level of filtering is increased. Thus the selection of K is a compromise between the need to reduce the random variation in occupancy and the requirement to closely track the real time trend.

Table 7.4 Example of metering rate schedule

Occupancy (%)	Metering rate (vehicles/min)
≤10	12
11–16	10
17–22	8
23–28	6
29–34	4
>34	3

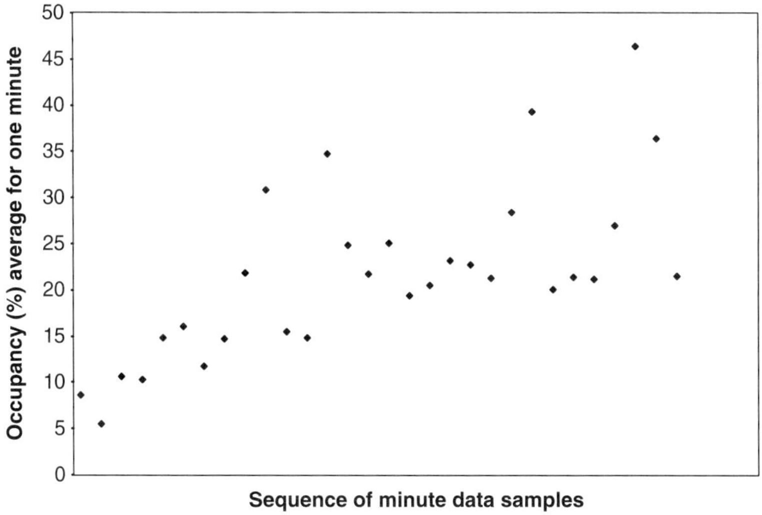

Fig. 7.12 Occupancy data vs. time for 1 min intervals. (Developed from Portland Oregon Regional Transportation Archive Listing – PORTAL)

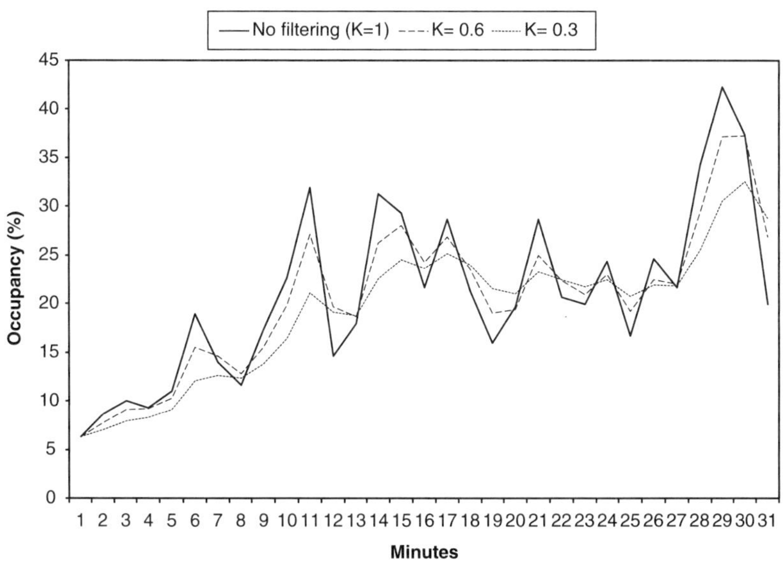

Fig. 7.13 Effect of filter coefficient on reported occupancy trend

Similar filtering considerations apply to the other traffic variables (e.g., volume and speed) that may be used for other purposes such as the geographical display of data in the traffic management center, use of data for CMS messages, and communication of this information to traffic service providers.

Other data filtering techniques such as Kalman filtering [14] have also been employed in traffic systems. Kalman filters automatically adjust the filter coefficients based on the random variation in the sensed value of occupancy and in the detection error.

7.4.3.3 Closed Loop Control

The technique described in Sect. 7.4.3.2 for scheduling metering rate by occupancy measurements does not control the metering rate to achieve a specific objective. The ALINEA strategy controls the metering rate to a desired level of occupancy based on the principles of linear control theory [15]. The desired level of occupancy (i.e., the occupancy set-point) may be established based on capacity or flow breakdown probability. The equation used for the metering rate computation is

$$R(j) = R(j-1) + K \bullet (\theta_S - \theta_M(j)) \tag{7.6}$$

where

$R(j)$ = meter rate (ramp volume) after time j
K_R = adjustable parameter
θ_S = occupancy set-point
$\theta_M(j)$ = occupancy measured by mainline detector (usually somewhat downstream of ramp merge with mainline) between time intervals $j-1$ and j

Figure 7.14 describes the operation of the control loop. The meter rate is added to the volume upstream of the meter (q_U). The model includes a physical time delay in the merged volume when sensed at the detector. This time delay is equal to the travel time between the ramp meter and the detector station. The detector station sensing this traffic provides the value of occupancy as provided by (7.3) and (7.4). As shown in (7.6), this occupancy θ_M is subtracted from the set-point occupancy value θ_S, and the difference is multiplied by parameter K_R. This product constitutes the change in metering rate from the prior computation interval. Because these changes are relatively small for each interval, the filtering of occupancy data as described in the previous section is not required. ALINEA has been deployed at a number of sites in Europe.

7.4.4 System-Wide Traffic-Responsive Restrictive Ramp Metering

System-wide traffic-responsive ramp meter strategies adjust the rates of ramp meters as a group to optimize some objective function. A commonly employed objective is to minimize the probability of demand exceeding capacity at any bottleneck location within the controlled section of roadway based on current sensed traffic conditions. This may entail reducing the metering rate at a number of ramps upstream of the bottleneck. Advantages of these strategies include the ability to respond to current changes in traffic demand, and the ability to respond to changes in roadway capacity resulting from weather conditions and incidents. A discussion

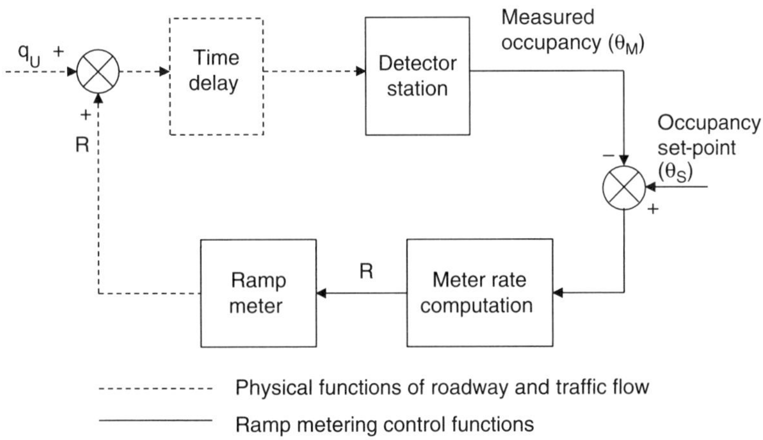

Fig. 7.14 ALINEA model

of a number of strategies is provided by Bogenberger and May [16]. Table 7.5 provides a summary of the key features of these strategies.

In general, these strategies examine detector data in a zone downstream of the meter. Occupancy or density from one or more critical downstream detector locations is compared with criteria (usually at or somewhat less than a value that represents capacity, or the variable's flow breakdown value). In some cases, density is computed by a count-in, count-out process for a zone.

The metering rates of one or more upstream ramp meters are established to adjust the occupancy or density values at the detectors so that the critical value is not exceeded. Many strategies incorporate a local metering algorithm in addition to the system-wide algorithm. The more restrictive rate computed by these algorithms is implemented.

The ramp is usually instrumented with one or more detector stations upstream of the meter in order to adjust the rates so that the queue does not spill back onto the surface street so as to obstruct traffic flow (this topic is discussed in Sect. 7.4.5.2).

7.4.5 Design Issues

7.4.5.1 Ramp Design Considerations

Metered ramps may be configured in a variety of ways to service signal display, safety, and storage requirements. The number of metered lanes and the use of single vehicle metering or platoon metering depends on the maximum metering rate to be employed (see Table 7.2). Where one lane is metered, vehicles may be stored in one lane or in two lanes and merged prior to the meter. In some cases an additional lane may be employed to permit buses or other high occupancy vehicles to by-pass the ramp meter. The California Department of Transportation (CALTRANS) provides a ramp meter design manual that recommends design criteria for metered ramps and provides examples of designs [17].

Table 7.5 Examples of system wide traffic-responsive ramp metering strategies

Strategy	Example of location used	References	Key properties of strategy
Stratified zone metering	Twin Cities, Minnesota	[24]	• Freeway divided into zones. Strategy sets metering rate to keep the number of vehicles entering the zone less than the number leaving • Zones are organized into groups (layers) that serve as the basis for meter rate calculations • Strong emphasis given to constraint on maximum ramp waiting times. Ramp queues are monitored to assure that waiting time is not exceeded
Fuzzy logic	Seattle, Washington	[25]	• See Appendix E
Helper	Denver, Colorado	[16]	• Local algorithm initially sets metering rates • If a ramp is metered at its minimum rate, the rates of upstream ramps are reduced
System wide adaptive ramp metering algorithm (SWARM)	Orange County, California	[26]	• Selects lower of SWARM 1 and SWARM 2 metering rates • SWARM 1 – Forecasts density and measures excess density at each detector station – Calculates target density and required volume reduction at each detector station – Assigns volume reduction to upstream meters • SWARM 2 – Local algorithm. Computes local density from headway measurements
Metaline	Paris	[27, 28]	• Generalizes ALINEA for system wide metering – Establishes an occupancy set-point for critical downstream detectors and measures the error between the set-points and the occupancy value – Provides incremental changes to each meter's metering rate based on weighted downstream occupancy error

Restrictive ramp metering often requires considerable storage space for vehicles. Caltrans describes a method for computing the ramp storage space required, Fig. 7.15 [17] consists of a grid divided horizontally into ten minutes time increments and vertically into 100 vph volume increments. The figure is based on an estimate of metering rates such as would be developed for a pretimed metering schedule. Hourly volumes (in hundreds) are entered into the first row of the table below the figure when the arrival rate exceeds the metering rate. The second row represents the discharge rate which is the metering rate for the period that the queue is present. The third row represents the queue in tens of vehicles and is computed as the difference between the first row and the second row plus the residual queue remaining from the preceding interval. Thus, in the example in Fig. 4.15, the maximum queue is computed as 60.4 vehicles, and occurs between 40 and 50 min after the demand first exceeds the metering rate as indicated by the vertical profile rate. Caltrans recommends that nine meters of ramp lane length be allocated for each vehicle stored, thus the minimum storage requirement is 540 lane meters of lane (60•9) in the example. Additional storage is recommended where there are significant percentages of trucks, buses, or recreational vehicles. The enclosed CD contains a data free image of this chart as shown in Fig. 7.16 it may be used to assist in establishing ramp metering schedules.

In many cases, ramps cannot accept or be modified to accept the storage requirement. When this occurs, it may be advisable to consider setting the metering rate to a higher value than that required to keep demand below capacity or to consider non-restrictive metering.

7.4.5.2 Control of Ramp Queue Length

During the course of a control period that uses restrictive ramp metering, the queue will build and wane in the general way illustrated by the vertical length in the shaded area of Fig. 7.15. Short term variations in this general profile result from random traffic arrivals or from vehicle platoons released by traffic signals upstream of the ramp. Most operating agencies limit the maximum queue length on the ramp for one or both of the following reasons:

- Extending the queue past the physical ramp or past the space set aside for the storage of vehicles destined for the ramp entry will interfere with general surface street traffic operations.
- Some agencies may choose to limit the ramp waiting time (see Sect. 7.5.2).

Thus, most ramp meter control systems provide a means for limiting the length of the queue, the technique usually utilizing data from the ramp queue detector shown in Fig. 7.2. The following describes a number of techniques that may be used to control the queue. The first four use the periodic occupancy measurement commonly developed by freeway management systems to determine the presence of the queue over the queue detector. Because the queue continues to build even after the metering control has been changed to compensate for the queue presence over the detector, that detector must be located somewhat downstream on the ramp from

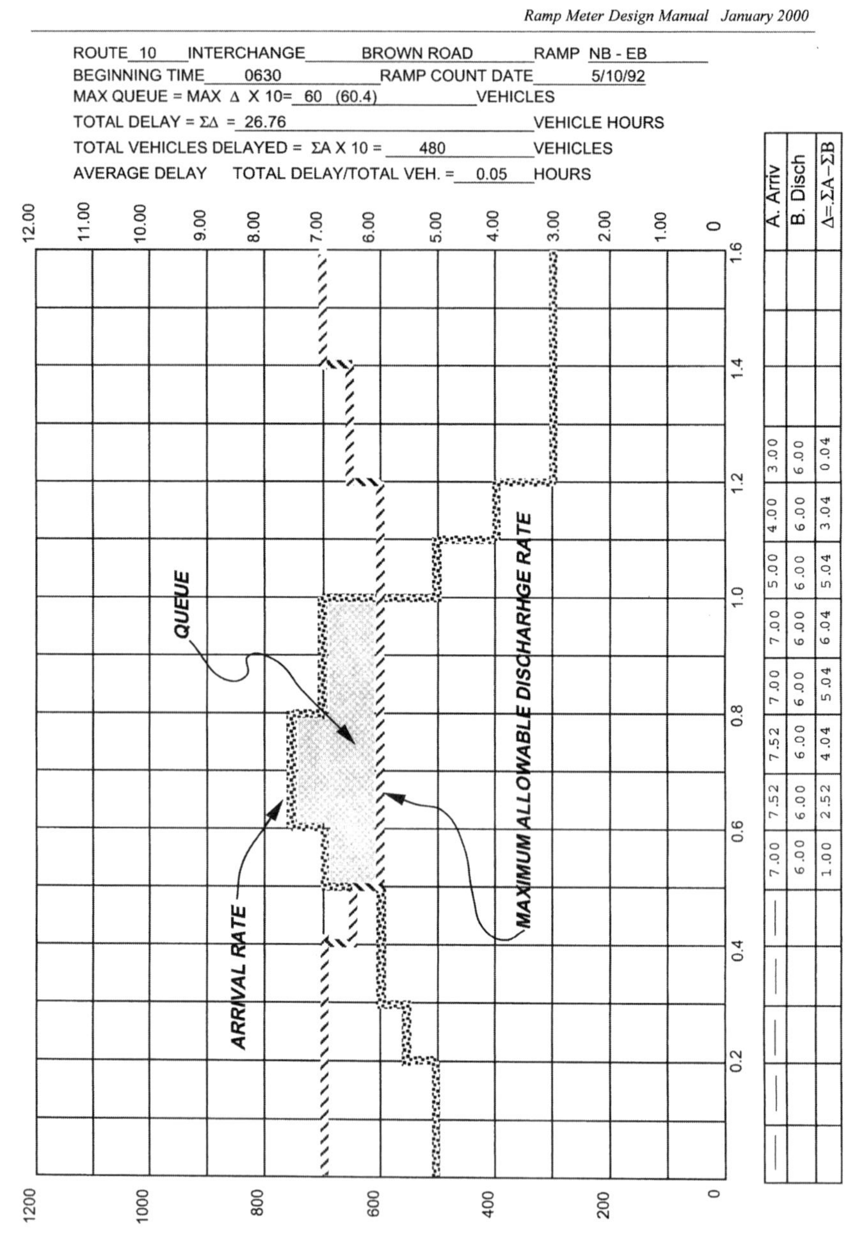

Fig. 7.15 Ramp queue computation chart

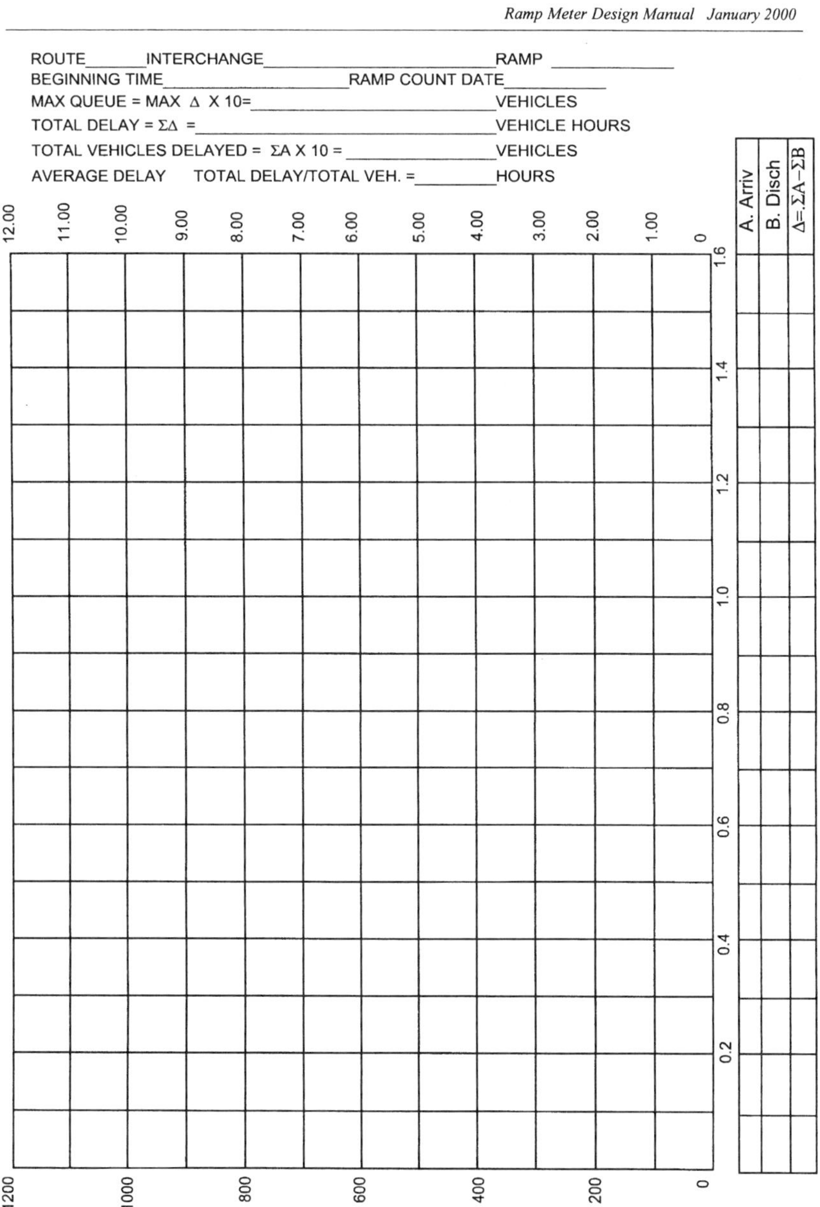

Fig. 7.16 Ramp queue computation chart (data free)

the maximum expected end of the queue. The fifth technique utilizes additional queue detector data from each vehicle to estimate the end of the queue and control the metering rate accordingly.

1. Termination of ramp metering when the queue is sensed by a high value of occupancy at the queue detector. Metering is resumed when the queue is no longer over the queue detector. This is an early approach that is still sometimes utilized. It is not recommended because termination of ramp metering has a severe adverse effect on mainline traffic flow.
2. Increase of ramp metering rate to a higher value than the ramp arrival rate when the queue is sensed by a high value of occupancy at the queue detector. The rate may be increased to the maximum ramp metering rate. The planned metering rate is restored when the queue is no longer over the queue detector. This type of control results in a limit cycle (oscillation of the queue) in the vicinity of the queue detector, requiring the detector to be placed upstream of the location to be protected from queue spillback. The effect is to reduce the available queue storage space on the ramp.
3. When the queue is sensed by a high value of occupancy at the queue detector the metering rate may be incremented. This incrementation process continues with each sampling period until the queue is no longer over the queue detector, at which time the planned metering rate is restored. This control process, used by many of the ramp meters in California is slower than Method 2, and may lead to instability in controlling the queue [18].
4. In order to minimize the queue buildup after detection, Gordon [19] describes a technique that uses a combination of a faster than conventional sampling period for the occupancy detector (10 seconds is recommended) in conjunction with a data processing technique that anticipates the presence of the queue at the location of the detector by including the rate of change of occupancy. This technique results in a limit cycle that has a lower amplitude when compared with Method 2, resulting in lower queue storage space requirements.
5. Sun and Horowitz [18] describe a technique that is applicable when the queue is close to the queue detector. At this point it adjusts the metering rate to maintain a prescribed queue length. Queue length is estimated by comparing the speed of each vehicle passing over the detector to a stored profile of vehicle speed vs. distance from the tail of the queue. This profile is developed beforehand based on observed measurements fitted to a curve that represents the vehicle's deceleration profile.

7.4.5.3 Freeway-to-Freeway Ramp Metering

Traffic patterns on a freeway in major metropolitan areas are often dominated by traffic merging from another freeway. Often the effect is to create excessive delays to motorists on the first freeway upstream of the merge point. To mitigate these delays, the ramp leading from the merging freeway may be metered to induce some motorists to enter the freeway at a downstream entry ramp. Jacobson and Landsman offer guidelines for the selection of appropriate sites [20]. These are summarized in Table 7.6.

7.5 Ramp Metering and the Motorist

7.5.1 *Motorist Benefits and Disbenefits Resulting from Ramp Metering*

A review of ramp metering projects in North America indicates that significant overall benefits for travel time in the corridor controlled by ramp meters and in the accident rate on the freeways controlled by ramp meters are obtained [8]. However, unlike other ITS treatments that are often characterized by benefits to all users, ramp metering, and especially restrictive ramp metering, provides benefits to some motorists and exacts penalties on others.

Beneficiaries of ramp metering generally include motorists whose ramp waiting time is short compared to their trip length on the freeway. In some cases, motorist may enter the freeway upstream of the metered section and experience no waiting time. Other beneficiaries may include transit passengers in buses that utilize a ramp meter by-pass lane. Motorists penalized by ramp metering include the following:

- Motorists whose ramp waiting time is relatively long compared to the travel time on the mainline. These motorists often enter the freeway in the central city closer to the central business district. In other cases, the waiting time is considerably different at different ramps. These equity issues have been addressed in the following ways:
  - Metering may be primarily utilized on suburban ramps rather than in the central city.

Table 7.6 Guidelines for freeway-to-freeway ramp metering [20]

- Consider locations where recurrent congestion is an issue or where route diversion should be encouraged
- Consider route diversion only where suitable alternative routes exist
- Avoid metering twice within a short distance
- Avoid metering single lane freeway-to-freeway ramps that feed traffic into an added lane
- Do not install meters on any freeway-to-freeway ramp unless analysis ensures that mainline flow will be improved so that freeway-to-freeway ramp users are rewarded
- Install meters on freeway-to-freeway ramps where more than one ramp merges together before feeding onto the mainline and congestion on the ramp occurs regularly (four or more times a week during the peak period)
- If traffic queues that impede mainline traffic develop on the upstream mainline because of a freeway-to-freeway ramp meter, then the metering rate should be increased to minimize the queues on the upstream mainline, or additional storage capacity should be provided
- Freeway-to-freeway ramp meters should be monitored and be controllable by the appropriate traffic management center
- Whenever possible, install meters at locations on roadways that are level or have a slight downgrade, so that heavy vehicles can easily accelerate. Also, install meters where the sight distance is adequate for drivers approaching the meter to see the queue in time to safely stop

  - Queue waits may be limited to an acceptable maximum (Section 7.5.2).
  - Metering may be implemented only in the outbound direction.

Section 3.2.1 discusses the Gini coefficient that may be used as a measure of equity.

- Motorists who divert from the freeway and motorists on the alternate routes whose trips do not utilize the freeway will generally experience longer trip times than they experienced before the freeway was metered. These issues may be mitigated in the following ways:
  - Preplanning by means of simulation or other techniques to provide metering rates that will limit diversion to a level acceptable to stakeholders concerned with surface street impacts. For example, when metering was first implemented in Portland, Oregon it was agreed to limit diversion so that surface street volumes would not increase by more than 25% [1].
  - Detectors in traffic signal systems and CCTV on surface streets may be monitored in real time, and metering rates adjusted to assure that an acceptable level of service is maintained on the alternates.

Issues such as these require the stakeholders to agree to the planned ramp metering operations and their estimated impacts. It is therefore recommended that ramp metering and its impacts be included in the Concept of Operations for the system (Sect. 2.1.2).

7.5.2 Public Acceptance of Ramp Metering

When ramps are restrictively metered, motorists waiting in the daily queue rapidly become aware of a delay that they had not previously experienced. The benefits in terms of reduced mainline travel time and more safe merges may not be as easily perceived. The motoring public, political leaders, and law enforcement agencies should be made aware of the potential benefits before they experience ramp metering. Public outreach techniques include brochures, use of the print and electronic media, and outreach to local leaders and law enforcement officials [1, 2]. In many cases these techniques have resulted in positive public attitudes towards and compliance with ramp meter signals.

As an example of the difficulty that acceptance of ramp metering may experience because of perceived long ramp waiting times, the Minnesota legislature in 1990 required a study to be conducted to assess the benefits of ramp metering at the behest of the public and political leaders. The study involved measuring delay with and without ramp metering, the ramp meters being turned off for a period of approximately 2 months.

The results showed that the ramp meters made considerable improvement in travel time reliability, mainline throughput and in crash reduction. A user survey indicated that while respondents now had an increased appreciation of the role of ramp meters, they felt that there was too much metering under free flow conditions and too many meters in general. The study recommended a new principle for ramp meter operation namely: *Balance the efficiency of moving as much traffic during the rush hours as possible, consistent with safety concerns and public consensus regarding the queue length at meters* [21].

As a result of the study, ramp metering policy was modified to limit ramp meter wait time to less than four minutes for local ramps and less than two minutes for freeway-to-freeway connector ramps. Implementation of this plan resulted in system-wide benefits that were better than the no metering situation, but worse that the previously employed ramp metering strategy [22].

7.6 Benefits Model for Ramp Metering

When demand is sufficiently close to capacity, ramp metering can reduce overall corridor delay and provide savings in fuel consumption and emissions in the affected roadway segment. It also reduces the accident rate near metered ramps. Figure 7.17 shows the relationships for the benefits model incorporated into the Design ITS model [23]. The model estimates annual delay and accident reduction in a segment of the highway influenced by ramp metering.

The symbols in the model are defined in Table 7.7.

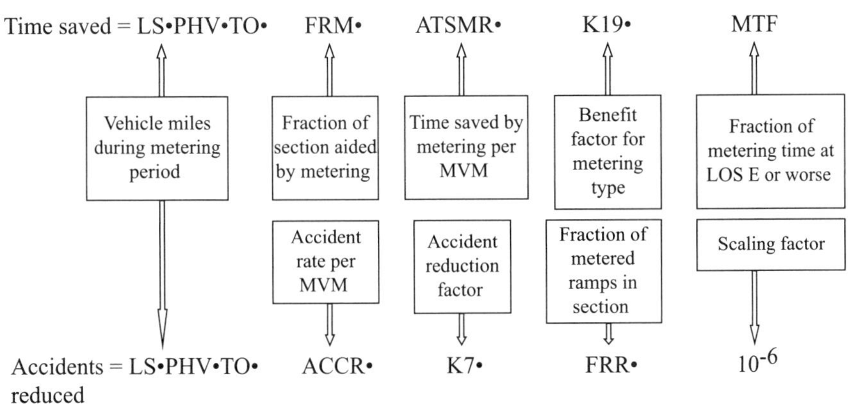

Fig. 7.16 Design ITS ramp metering benefits model

Table 7.7 Symbols for ramp metering benefits model in Fig. 7.16

Symbol	Definition	Design ITS default value
ACCR	Accident rate for section	Default not applicable
ATSMR	Average time per mile per vehicle saved by metering	0.00476 h per mile per vehicle
FRM	Fraction of roadway segment improved by metering	Default not applicable
FRR	Fraction of ramps in roadway segment that contain ramp meters	Default not applicable
K7	Accident reduction factor	0.2
K19	Ramp metering benefit factor	K19 = 1.0 for restrictive metering K19 = 0.4 for nonrestrictive metering
LS	Length of roadway segment influenced by ramp metering (miles)	Default not applicable
MTF	Fraction of planned metering period that freeway is at level of service E or worse	Default not applicable
PHV	Peak hour volume	Default not applicable
TO	Metering hours per year	Default not applicable

References

1. Neudorff LG et al (2003) Freeway management and operations handbook. Report FHWA-OP-04-003, Federal Highway Administration, Washington, DC
2. Jacobson L et al (2006) Ramp management and control handbook. FHWA-HOP-06-001, Federal Highway Administration, Washington, DC
3. Gazis D, Foote R (1967) Surveillance and control of tunnel traffic by an on-line digital computer. 32nd National Meeting Operations Research Soc. Am
4. Persaud B et al (2001) Study of breakdown-related capacity for a freeway with ramp metering. Transportation Research Record 1748:110–115, Transportation Research Board, Washington, DC
5. Banks J (2002) Review of empirical research on congested freeway flow. Transportation Research Record 1802:225–232, Transportation Research Board, Washington, DC
6. Shawky M, Nakamura H (2007) Characteristics of breakdown phenomena in Japan urban expressway merging sections. Transportation Research Record 2012:11–19. Transportation Research Board, Washington, DC
7. Zhang L, Levinson D (2004) Ramp metering and the capacity of active freeway bottlenecks. 83rd annual meeting of the Transportation Research Board, Washington, DC
8. Piotrowics G, Robinson J (1995) Ramp metering status in North America – 1995 update. Federal Highway Administration, Washington, DC
9. Twin Cities ramp meter evaluation (2001) Cambridge Systematics
10. May AD (1990) Traffic flow fundamentals. Prentice Hall, Englewood Cliffs
11. Imada T, May AD (1985) FREQ8PE – a freeway corridor simulation and ramp metering optimization model. Report UCB-ITS-RR-85-10, University of California, Berkely, CA
12. Blumentritt CW et al (1981) Guidelines for selection of ramp control systems. NCHRP Report 232, Washington, DC
13. Gordon RL et al (1995) Traffic control systems handbook. Report FHWA-SA-95-032, Federal Highway Administration, Washington, DC
14. Gelb A et al (1974) Applied optimal estimation. MIT Press, Cambridge MA
15. Papageorgiou M et al (1991) ALINEA: a local feedback control law for on-ramp metering. Transportation Research Record 1320:58–64

16. Bogenberger K, May AD (1999) Advanced coordinated traffic-responsive ramp metering strategies. Report UCB-ITS-PWP-00-19, University of California, Berkeley, CA
17. Ramp meter design manual (2000) Traffic Operations Program, California Department of Transportation, Sacramento, CA
18. Sun X, Horowitz R (2005) Localized switching ramp-metering control with queue length estimation and regulation and microscopic simulation results, Proceedings of the 2005 American Control Conference, vol 3, pp 2141–2146
19. Gordon RL (1996) Algorithm for controlling spillback from ramp meters. Transportation Research Record 1554, Transportation Research Board, Washington, DC
20. Jacobson EL, Landsman J (1994) Case studies of freeway to freeway ramp and mainline metering in the U.S. and suggested policies for Washington State. Transportation Research Record 1446:48–55. Transportation Research Board, Washington, DC
21. Executive summary – Twin Cities ramp meter evaluation (2001) Cambridge Systematics
22. Executive summary – Twin Cities ramp meter evaluation – Phase II interim ramp meter strategy, Phase III Plans for new ramp meter strategy (2001) Cambridge Systematics
23. Gordon RL (2007) Design ITS user's manual
24. Stratified zone metering – The Minnesota algorithm (ND) Minnesota Department of Transportation
25. Taylor C et al (2000) Results of the on-line implementation and testing of a fuzzy logic ramp metering algorithm. 79th Annual Meeting of the Transportation Research Board, Washington, DC
26. System wide adaptive ramp metering algorithm – high level design (1996) National Engineering Technology Corporation
27. Papageorgiou M et al (1990) Modeling and real time control of traffic flow on the southern part of the Boulevard Peripherique in Paris: Part I Modeling. Transport Res 24A(5):345–359
28. Papageorgiou M et al (1990) Modeling and real time control of traffic flow on the southern part of the Boulevard Peripherique in Paris: Part II Coordinated on-ramp metering. Transport Res 24A(5):361–370

Chapter 8
Communications for ITS

Abstract This chapter describes the communication standards, system design considerations, and the general features of the technologies required for communicating between the transportation management center and the field devices in a reliable and cost effective manner. The chapter discusses:

The Open System Interconnection Architecture (OSI) and the National Transportation Communications for ITS Protocol (NTCIP) communication standards.

Types of communication channels and communications backbone and distribution systems.

Communication technologies at an appropriate level for ITS concept development. An example of communications system design at this level is described.

8.1 Introduction

Communications among transportation management centers (TMCs) and between TMCs and field devices constitute a key and often expensive ITS function. The key challenge for the ITS communications system designer is to provide the means for the TMC to communicate with ITS field devices in a cost-effective way. The Telecommunications Handbook for Transportation Professionals [1] describes ITS communication technologies and their application in detail. This chapter provides a general overview of the communications related issues that are important to ITS concept design and for preliminary cost estimation purposes. Detailed communication design trade-offs and final design decisions should be performed by an experienced ITS communications engineer.

Topics discussed in this chapter are:

- Communication standards for freeway ITS.
- Channelized communications.
- Backbone and distribution systems.
- Communications technologies and their relative attributes.
- An example of a high-level communication architecture and the basis for its design.

R. Gordon, *Intelligent Freeway Transportation Systems: Functional Design*,
DOI 10.1007/978-1-4419-0733-2_8, © Springer Science+Business Media, LLC 2009

8.2 Communication Standards

Communication requirements include supporting functions that are derived from the particular data to be transmitted to the physical devices employed for communication. The Open System Interconnection (OSI) Model is an internationally recognized seven layer structure that provides services to the layer above it and receives services from the layer below it. Table 8.1 shows this structure [2].

In earlier years it was common for ITS to employ proprietary communication protocols, and in some cases this practice continues. The use of proprietary communications equipment limits the selection of suppliers for equipment replacement and system expansion purposes, and makes it difficult to change or expand the communications system without affecting the compatibility of existing field devices. To provide for *interoperability* (support of different ITS devices on a common communication channel) and *interchangeability* (ability to use devices from multiple manufacturers on the same communication channel) a set of national standards was developed. The standard set, based on the OSI model, is termed the *National Transportation Communications for ITS Protocol* (NTCIP). The ITS standards and

Table 8.1 Network communications through OSI model

Layer	Properties
7 Applications	Provides services directly to user applications. Because of the potentially wide variety of applications, this layer must provide a wealth of services. Among these are services establishing privacy mechanisms, authenticating the intended communication partners, and determining if adequate resources are present
6 Presentation	Performs data transformations to provide a common interface for user applications, including services such as reformatting, data compression, and encryption
5 Session	Establishes, manages, and ends user connections, and manages the interaction between end systems. Services include such things as establishing communications as full or half duplex and grouping data
4 Transport	Insulates the three upper layers, 5 through 7, from having to deal with the complexities of Layers 1 through 3 by providing the functions necessary to guarantee a reliable network link. Among other functions, this layer provides error recovery and flow control between the two end points of the network connection
3 Network	Establishes, maintains and terminates network connections. Among other functions, standards define how data routing and relaying are handled
2 Data-Link	Ensures the reliability of the physical link established at Layer 1. Standards define how data frames are recognized and provide necessary flow control and error handling at the frame level
1 Physical	Controls transmission of the bitstream over the transmission medium. Standards for this layer define such parameters as the amount of signal voltage swing, the duration of the voltages (bits), and so on

Courtesy of Novell, Inc

their applications are described in the NTCIP Guide [3] which identifies ITS standards for Layers 1–4 and for Layer 7. As the message developed for transmission descends through the protocol layer stack, additional data in the form of headers and footers are added to the previous layer. This *overhead* is often substantial, and adds considerably to the quantity of communication data that are actually transmitted. The tradeoff, using standards to achieve interoperability and interchangeability, is generally considered to be advantageous.

With minor exceptions, NTCIP Layers 1 through 4 and Layer 7 employ standards that are commonly used for data communication. NTCIP's unique contribution is to define standards for an *information level* that resides above the application level. This level provides *data dictionaries* (definition of data formats) and *message sets* (comprised of data dictionary elements) that differentiate ITS applications from others. The data dictionaries and message sets support center-to-center and center-to-field device communications.

Depending on the functions of the TMC and the agencies with which the TMC communicates, other standards may also be employed. These most commonly include:

- IEEE 1512 Standards [4]. This family of standards is often used by incident management centers and the TMCs that communicate with them. These standards provide message sets related to traffic management, public safety, hazardous materials, and incident response.
- Transit Communications Interface Profiles (TCIP) Standards [5]. This standards family emphasizes communication requirements for transit vehicles, between transit vehicles and fixed sites and at fixed site facilities.

8.3 General Communication System Design Considerations

A communication channel is a path, medium, or technology that enables information to be transferred from a sender to a receiver. The types of communication channels used for ITS applications include:

- Channels that are owned and operated by the operating agency including wireline and wireless technologies.
- Public network based communication services. These services may utilize:
  - Wireline technologies (usually leased on a per month pay schedule). Implementation of these services may require the operating agency to supply connecting cable from the service provider's termination point to the field device.
  - Cellular telephone based data services (usually charged on a quantity of data basis). This type of service is often less expensive to physically access than leased wireline services.
- Center-to-center communications are usually provided by public based networks. In some cases, government owned networks may be used.

Owned communication systems are often classified as *backbone* (or trunk) systems and *distribution* systems. A backbone system transfers network data at high speeds. Network backbones are designed to maximize the reliability and performance of large-scale, long-distance data communications. Large-scale ITS often feature a backbone system comprising fiber optic cable (fiber optic communications and other communication technologies are discussed in Sect. 8.4) and wireless channels. Backbone systems contain *nodes* or *hubs* that connect to other nodes and that transfer data to distribution systems. These systems carry the data from a backbone node or tie point to the field device. The distribution systems may use media or technologies that differ from those used for the backbone. Several field devices may be controlled on a single distribution system channel by means of communications multiplexing. An example of a multiplexing technique is the time sharing of communication with each field device. Multiplex techniques are also used for backbone communications.

Figure 8.1 illustrates a simple example of a backbone and the set of distribution systems attached to backbone Node 3. CCTV camera 1 is co-located with Node 3. This camera is located in a large interchange. Unlicensed microwave communications connect other devices in the interchange to this node. Unlicensed microwave links connect detectors 2 and 3 to the location of CCTV camera 2. All of this data is returned to node 3 through an unlicensed microwave link. Detector D1 and CMS 1 are also connected to Node 3 through separate unlicensed microwave links.

Most ITS field devices including traffic detectors, CMS, and highway advisory radio (HAR) require low data rate communication channels. High data rates, however, are needed to provide CCTV service of a quality that is acceptable to most system operators. Video communication requirements therefore often dominate the communication architecture requirements for ITS. Center-to-center communication

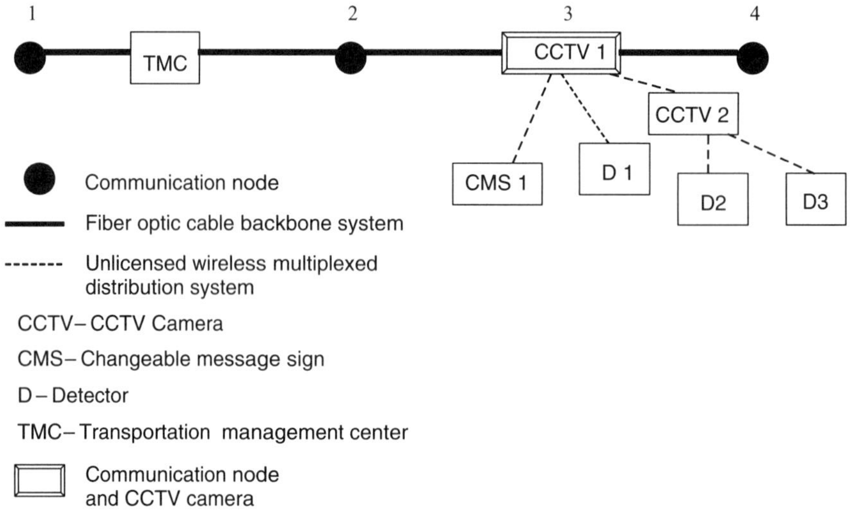

Fig. 8.1 Example of a communication system architecture

and communication with the public use additional methodologies and media such as the internet, conventional dial up telephone service, commercial television, and commercial AM, FM, and satellite radio.

8.4 Communication System Technologies for Center-to-Field Communication

This section presents an overview of the technologies that are most commonly used for freeway center-to-field communication.

8.4.1 Wireline Based Communications

Fiber optic cable is the principal wireline technology currently being installed for freeway ITS applications. Copper cable is sometimes used. Coaxial cable has been used in the past and is currently still in use for some ITS.

Fiber optic (or "optical fiber") refers to the medium and the technology associated with the transmission of information as light impulses along a strand of glass. A fiber optic strand carries much more information than conventional copper wire and is far less subject to electromagnetic interference (EMI). Transmission over fiber optic strands requires repetetion (or regeneration) at varying intervals.

Freeway ITS generally use single-mode optical fiber. A typical example of a single-mode fiber is shown in Fig. 8.2. This fiber is very thin (e.g., 8 μm). With single-mode fiber optic cable, the combination of the transmitter wavelength, the optical properties of the fiber cable core and the narrow diameter of the fiber core results in the propagation of light parallel to the axis of the cable. In multimode fiber cable the light is reflected from the cladding as it travels down the fiber resulting in greater signal loss or *attenuation*. While somewhat more expensive than multimode fiber, single-mode fiber can transmit signals at longer distances and at higher data rates. A number of fiber strands are generally bundled into a fiber optic cable.

Light emitting diodes or laser diodes convert electrical signals into optical signals and serve as transmitters. Photodiodes convert the light back into electrical signals and are used as receivers. Signals may be received at drop points along the cable by means of drop and insert units. These units allow the information to be received at the drop point, allow information developed at the drop point to be transmitted to a local user, and serve to regenerate the strength of the signal for transmission to the next drop or insert unit.

The basically high reliability of fiber optics communications may be further enhanced by connecting the fiber in a protected ring configuration, an example of which is shown for the backbone system in Fig. 8.3. Information is transmitted around the ring in both directions (this may be accomplished by using two fibers or alternatively on one fiber using transmissions in opposing directions on two different

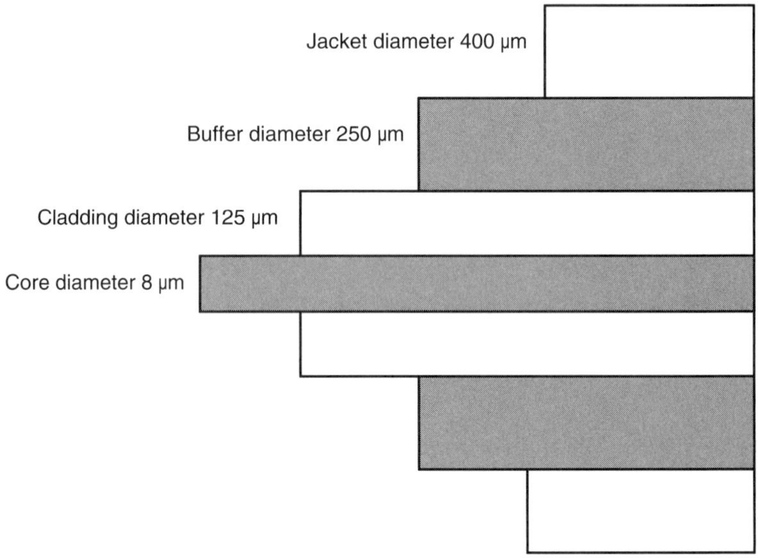

Fig. 8.2 Typical single-mode fiber optic cable cross section

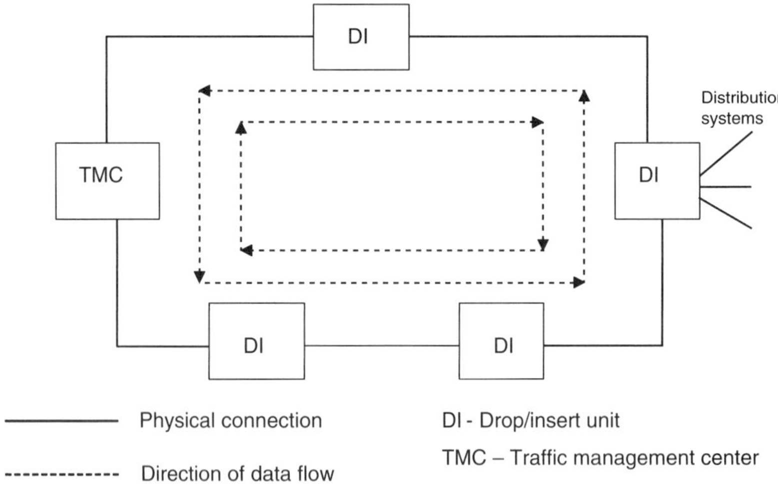

Fig. 8.3 Protected ring configuration

wavelengths). If a drop or insert unit fails, only the information to those distribution networks associated with the unit is lost. If the fiber network includes a set of roadways that form a geometric loop, then a cut in the cable will not affect data transmission.

The capital cost of a fiber optics cable installation is relatively high on a per mile basis. Since the principal installation cost results from the trenching and burial of conduit, the incremental cost of adding more fibers is relatively low. Coupled with its high reliability, data carrying capacity and freedom from interference, it is most cost effective when used for ITS that feature a dense equipment deployment (relatively large number of devices per mile). Since it is relatively inexpensive to install fiber optic cable when the roadway is being initially constructed, or major reconstruction is being performed, provisions for a conduit system for fiber optic cable are often included in these projects.

8.4.2 Wireless Communications

Some ITS use microwave technologies that are licensed by the Federal Communications Commission as well as unlicensed microwave technologies. Licensed microwave provides for dedicated use of a radio spectrum channel between two points at prescribed transmitter power levels. This channel is secure from competition by other potential users. Freeway management systems typically employ channels in the 18 and 23 GHz frequency bands. Licenses may be difficult to obtain in large metropolitan areas with high population densities. A line of sight between the terminations of the communication link is required. Transmission at these frequencies may be subject to interruption or reduced performance during heavy rainfall, particularly for the longer transmission links. Transmission in the 80 GHz band is also used for some ITS, however, rainfall interference is more severe in this frequency band.

Although equipment and licensing costs for these systems are significant, the total installation cost is generally less than that for fiber optic cable systems. By using two transmitters and receivers with different *polarization* orientations[1] a protected ring configuration for a licensed microwave communication system can be provided (at additional cost).

Unlicensed microwave systems, while relatively inexpensive, are subject to interference by other users. The interference is often severe in major metropolitan areas and increases with the length of the transmission link (the longer the link, the greater the opportunity for other users in the area to interfere with the signal). Equipment using this technology (spread spectrum radio) seeks to minimize this interference by searching for appropriate channels and by using various modulation schemes that are less sensitive to background interference. The use of such techniques may reduce the data transmission rate. The data rate reduction is less serious for low speed data devices than for CCTV transmission, where image quality may be seriously affected. The unlicensed frequencies generally used for ITS are in the 900 Hz band for low data rate applications. Examples for higher bandwidth for CCTV applications or combined CCTV and low data rate applications include the 2.4, 4.3, and 5.8 GHz bands. Figure 8.4 shows a microwave detector (sensor) with an integrated wireless communication capability.

[1] Polarization refers to the direction of the electromagnetic wave's electric field.

Fig. 8.4 Remote traffic microwave sensor (RTMS) with integrated wireless link. Courtesy of Image Sensing Systems, Canada Ltd

8.4.3 Public Network Based Communication Services

The most commonly used public network communication services currently employed include cellular telephone data services and leased telephone cable. Pager services are sometimes employed for communication with HAR beacons.

8.4.3.1 Cellular Telephone Data Services

Cellular telephone carriers generally provide data services based either on CDMA (code division multiple access) or GSM (global system for mobile communications). These technologies provide the capability for short message services (text messaging). Cellular telephone data services are particularly useful to communicate with low speed ITS devices such as CMS, HAR, and traffic detectors. Their cost effectiveness improves when the devices are located at a considerable distance (e.g. several miles) from a backbone communication system or from the TMC. Charges for the service are based on the quantity of data transmitted. The cost usually includes a fixed monthly cost for a prescribed quantity of data and an additional cost for data in excess of this quantity. While this quantity is generally sufficient to service the low data speed devices, communication of CCTV signals generally requires higher data transmission rates. Therefore the total amount of data required to provide images of acceptable quality to most system operators significantly increases the cost. Sometimes system operators will use a lower data rate (which provides a lower frame repetition rate and lower resolution) but will switch to a higher data rate when conditions such as incidents warrant. The capital cost for installations in the field and at the TMC are low.

During periods of high usage by other customers, public cellular telephone services are subject to difficulty in connecting to the system and in providing full data rate service. Service reliability is an issue to consider during system design.

8.4.3.2 Leased Communication Channels

Communication channels may be leased from a communications service provider, often the local telephone company (TELCO.) These channels usually consist of copper or fiber optic cable from a TELCO central office to the TMC and from a TELCO central office to a termination point near the field device. The operating agency is responsible for connecting from this point to the field device (see Fig. 8.5). Communication among TELCO central offices may be provided by TELCO selected media. The quality of service from the TMC to the drop point is the responsibility of the service provider.

Service for low data rate devices may be provided by a leased voice-grade channel, while service for CCTV cameras is generally provided by a T1 channel or fractional T1 channel[2]. Since the user leases a dedicated channel, there is no competition from other users (as is the case with public cellular service) and the service is therefore more reliable. Service cost is usually computed as the sum of the following:

- Charges from a TELCO central office to the TMC.
- Charges from a TELCO central office to the field drop.
- Charges based on the straight line distance from the TELCO central office serving the TMC to the TELCO central office serving the field drop.

Table 8.2 compares cellular and leased channel public network based communication.

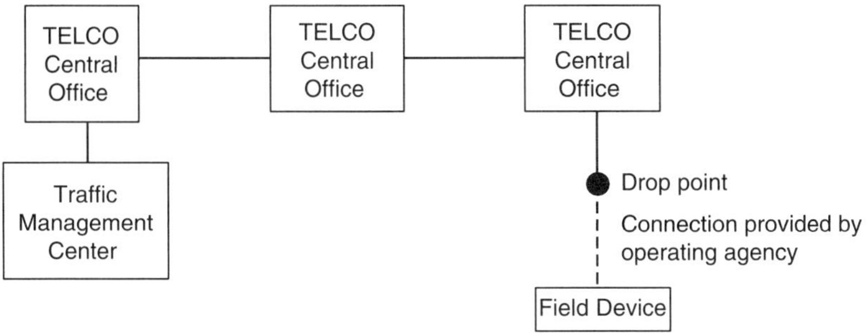

Fig. 8.5 Typical connections using communication service provider

[2]A T1 channel provides service rates of 1.544 Mbps (megabits/second). While this service is often provided by telephone companies, the area for which service is available may be less than the coverage area for voice-grade service. The leasing charge for this service is generally relatively high.

Table 8.2 Comparison of cellular and leased channel public network based communications

Issue	Cellular	Leased channel
Reliability of service	May be subject to difficulty in connecting and maintaining data rate during periods of high usage by other customers	Reliable
Capital cost	Low	Higher because service from service provider drop point to field devices must be provided and maintained by the operating agency
Operating cost for low data rate service	Low and insensitive to distance from field device to TMC	Generally higher than cellular. Depends on distance from field device to TMC
Operating cost for CCTV	Generally high for high quality of service. Quality of service may be varied by system operator as circumstances require	Generally high and depends on distance from field device to TMC. Cost vs. picture quality may be traded off by selection of level of fractional T1 service. Service level not changeable by TMC operator

8.5 Example of Communications Concept Design

This section describes a conceptual scenario for a communications design alternative using the technologies discussed in Sect. 8.4.

An ITS is to be installed in a metropolitan area with a population of approximately 400,000. ITS installation will be performed concurrently with the reconstruction of a section of freeway. A site for the TMC that is close to the freeway has been selected. The region exhibits moderate rainfall conditions. Current users of unlicensed wireless communication in the area report that very reliable service is experienced at link distances of up to one mile for low data rate communications and at distances ranging from one-half mile to one mile for video. Service at somewhat longer ranges is mostly reliable, but interference is experienced at certain locations and at certain times.

Since fidelity of communication service and project cost are often highly related, the following general guidelines were established for the project:

- Unlicensed microwave radio may be used for links servicing low speed devices for link distances up to one mile, and for CCTV for link distances up to one-half mile.
- Unlicensed microwave radio may be used where a few devices (e.g., three) are serviced by the link. The number of devices on the link may be considered as a rough measure of the impact on the system to the loss of communication or the degradation of communication quality on the link.[3]

[3] The acceptable number of devices on a link for a technology varies with the designer's assessment of the acceptable level of potential communication quality degradation. The values in this example reflect the author's assessment.

- Licensed microwave radio without a protected loop may be used for links where several devices (e.g., up to ten) are serviced by the link.
- Licensed microwave radio with a protected loop may be used for links where many devices (e.g., more than ten) are serviced by the link.

The preliminary field equipment deployment plan was developed to satisfy functional project requirements. Based on this, a preliminary communications plan was generated as a basis for estimating project cost during the project scoping and early design phases, and to serve as a strawman[4] for detailed future communication design studies. This plan is shown in Fig. 8.6. The technologies to implement the communication links in the plan are shown in the in the figure. Link distances are not shown to scale. The device identification numbers in the figure are used in the development of the plan (Table 8.3). Alternative plans may also merit consideration.

The table shows the characteristics of the communication links. The second column in the table provides the length of the link and the third column identifies the number of field devices serviced by the link. The fourth column shows the

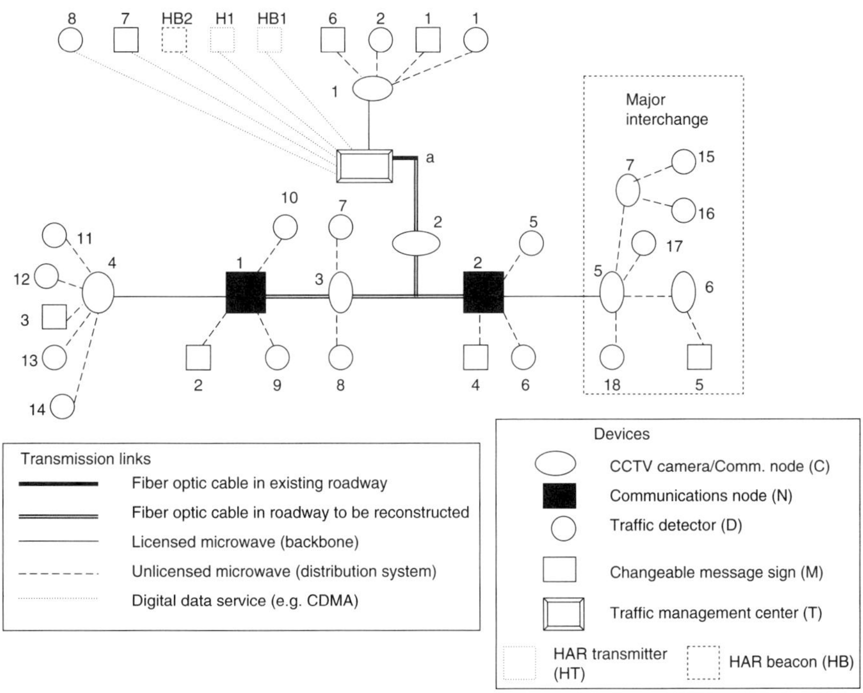

Fig. 8.6 Preliminary plan for center to field communication

[4] A strawman is a plan or document that serves as the starting point in the evolution of a project. The strawman is not expected to be the last word; it is refined until a final model or document that resolves all issues is obtained.

Table 8.3 Characteristics of communication links

Link ID	Link length miles	Number of devices on link	Link technology	Basis for technology selection
T-a	0.1	25	Fiber optics in existing roadway	Short link – very high data reliability needed
a-C2	0.3	25	Fiber optics in reconstructed roadway	High reliability at moderate cost
C2-C3	0.8	12	Fiber optics in reconstructed roadway	High reliability at moderate cost
C3-S7	0.3	1	Unlicensed wireless	Short link – few devices
C3-S8	0.3	1	Unlicensed wireless	Short link – few devices
C3-N1	0.5	9	Fiber optics in reconstructed roadway	High reliability at moderate cost
N1-S9	0.4	1	Unlicensed wireless	Short link – few devices
N1-S10	0.4	1	Unlicensed wireless	Short link – few devices
N1-M2	0.3	1	Unlicensed wireless	Short link – few devices
N1-C4	1.6	6	Licensed wireless	Medium length link – several devices
C4-S11	0.5	1	Unlicensed wireless	Short link – few devices
C4-S12	0.5	1	Unlicensed wireless	Short link – few devices
C4-S13	0.2	1	Unlicensed wireless	Short link – few devices
C4-S14	0.2	1	Unlicensed wireless	Short link – few devices
C4-M3	0.4	1	Unlicensed wireless	Short link – few devices
C2-N2	0.6	11	Fiber optics in reconstructed roadway	High reliability at moderate cost
N2-S5	0.3	1	Unlicensed wireless	Short link – few devices
N2-S6	0.6	1	Unlicensed wireless	Short link – few devices
N2-M4	0.5	1	Unlicensed wireless	Short link – few devices
N2-C5	2.1	8	Licensed wireless	Medium length link – several devices
C5-S18	0.2	1	Unlicensed wireless	Short link – few devices
C5-S17	0.2	1	Unlicensed wireless	Short link – few devices
C5-C7	0.7	3	Unlicensed wireless	Short link – few devices
C7-S15	0.2	1	Unlicensed wireless	Short link – few devices
C7-S16	0.2	1	Unlicensed wireless	Short link – few devices
C5-C6	0.8	1	Unlicensed wireless	Short link – few devices
C6-M5	0.3	1	Unlicensed wireless	Short link – few devices

T-H1	4.0	1	CDMA	Long link – less expensive than licensed microwave
T-HB1	6.0	1	CDMA	Long link – less expensive than licensed microwave
T-HB2	2.5	1	CDMA	Long link – less expensive than licensed microwave
T-S8	4.0	1	CDMA	Long link – less expensive than licensed microwave
T-M7	4.5	1	CDMA	Long link – less expensive than licensed microwave
T-C1	1.3	5	Licensed wireless	Medium length link – several devices
C1-M6	O.6	1	Unlicensed wireless	Short link – few devices
C1-S2	0.2	1	Unlicensed wireless	Short link – few devices
C1-M1	0.4	1	Unlicensed wireless	Short link – few devices
C1-S1	0.2	1	Unlicensed wireless	Short link – few devices

Symbols in Table
a – Change in fiber optics installation type from existing roadway to reconstructed roadway
C – CCTV Camera and communications node
H – Highway advisory radio transmitter
HB – Highway advisory radio beacon
M – Changeable message sign
N – Communications node
S – Traffic detector station
T – Traffic Management Center

technology selected. The rationale for this selection, shown in the fifth column is based on the expected vulnerability of the link to service interruption, and the life cycle cost of candidate technologies that can service link requirements.

References

1. Leader S (2004) Telecommunications Handbook for Transportation Professionals – The Basics of Telecommunications. Report No. FHWA-HOP-04-034, Federal Highway Administration, Washington, DC
2. Novell's Networking Primer, Novell, Inc.
3. The NTCIP Guide – Updated Version 3 (2002) American Association of State Highway and Transportation Officials, Institute of Transportation Engineers, National Electrical Manufacturers Association.
4. Ogden MA (2004) Guide to the IEEE 1512 family of standards. Institute of Electrical and Electronic Engineers, New York, NY
5. APTA TCIP – S – 3.0.0 Standard for Transit Communications Interface Profiles, American Public Transit Association, Washington, DC

Chapter 9
Transportation Management Centers

Abstract This chapter describes the functions of a transportation management center (TMC). Information flows among stakeholders and how they relate to the Regional ITS Architecture and how they may be implemented are discussed The functions of a typical major TMC are described.

9.1 Transportation Management Center Functions

The control and management of ITS operations is performed in a transportation management center (TMC), sometimes called a traffic management center or traffic operations center. The TMC provides a focal point for implementing the Regional ITS Architecture and the Concept of Operations (CONOPS) [1] developed from the project's systems engineering processes (Sect. 2.1.2). TMCs often house operations for a number of transportation related agencies and emergency service providers.

TMCs provide general management services to support transportation related functions provided by agencies responsible for freeway operations, surface street operations, emergency management, and police services. TMCs facilitate inter-agency communication and coordination, implement the provision of traffic information to the media and to the public, coordinate with transit agencies and provide a point of contact for the public and for organizations with special information needs. Major freeway management functions supported by the TMC are described below.

9.1.1 Support of Emergency Management Services

TMCs support emergency management service providers that respond to traffic related incidents. They make information and traffic conditions available to responders and help them identify the need for emergency services. They may identify the quickest route for emergency service providers to reach the incident site. Specific information such as CCTV images, lanes affected and the location of the tail of the queue may be afforded in response to requests by emergency service providers. The TMC

R. Gordon, *Intelligent Freeway Transportation Systems: Functional Design,*
DOI 10.1007/978-1-4419-0733-2_9, © Springer Science+Business Media, LLC 2009

generates, stores, and implements incident management plans and provides the incident support services discussed in Sect. 4.5.2.

CCTV is a key tool to assist in incident management as it supports the following functions:

- Identifying the types of response services required.
- Assisting responders in finding the quickest route to the incident.
- Assisting responders in managing traffic in the vicinity of the incident and its queue.

CCTV images are often shared with other TMCs, agencies, and emergency responders.

9.1.2 *Provision of Information to Motorists*

The TMC provides information to the agency's field devices, such as changeable message signs, highway advisory radio, HAR beacons, and perhaps kiosks in agency operated locations or private venues. Traffic information is also offered by private agencies such as independent service providers and the media. The TMC may provide messages requested by other TMCs. Information generated by the TMC typically makes a significant contribution to the information base for the state's 511 motorist information service.

The types of information provided to the motorist by the TMC may include:

- Nonrecurrent congestion and related problems such as police activity and roadway surface related problems.
- Recurrent congestion in the form of travel time or travel movement information (depending on the agency's policy).
- Construction scheduling and traffic impact.
- Special event scheduling and traffic impact.
- Major events or incidents on other facilities.
- AMBER alerts.
- Roadway or traffic conditions related to weather events.
- Transit and corridor related information including park-and-ride facility status.
- Default messages related to motorist safety (depending on the agency's policy).

9.1.3 *Operation of Ramp Meters*

TMCs support the following operations associated with ramp meters:

- Daily operations – Operating personnel select metering modes and may alter metering rates based on observations of mainline traffic flow and ramp queues. Incident management plans and corridor management functions may also require the modification of ramp metering rates.

- Data archiving and mining – Traffic detector data is stored and analyzed by computer programs. These programs provide information showing the distribution of traffic data at site by time (e.g., Fig. 6.3). Daily profiles showing speed for a section of roadway such as those displayed in Fig. 9.1 [2] and the volume data that accompany them are useful in determining and revising meter rates, schedules, and traffic responsive metering parameters. Data archiving and mining also support other TMC functions.
- Revision of metering plans – Meter rates, schedules, and traffic responsive metering parameters may be modified based on archived data as described above. Revisions may be facilitated by the use of traffic simulation programs [3].

9.1.4 Operation of Service Patrols

Freeway service patrols, also known as motorist assistance patrols and courtesy patrols may be operated by a highway agency or a private company. The TMC usually coordinates and supervises service patrol operations. Most service patrol vehicles follow pre-established *beats* or routes. Their primary function is to mitigate congestion or potential congestion by clearing minor incidents from travel lanes. They generally provide the following services:

- Assist immobile vehicles by providing fluids, performing minor mechanical repairs, and calling towing services. Push bumpers may be used to clear stalled vehicles from travel lanes.
- Clear debris from the moving lanes and from the shoulders.
- Assist police in the tagging and removal of abandoned vehicles.
- Assist response providers in the management of traffic during incidents.

The benefits provided by service patrols to the freeway system include reduction of delay and reduction in secondary accidents by improving the clearance time for incidents. They are popular with the motoring public.

9.1.5 Coordination of Traffic Signal Operation with Freeway and Corridor Requirements

Implementation of traffic diversion plans in the event of incidents may require the TMC to instigate alternative signal timing plans. In some cases, another agency implements the timing plans requested by the freeway TMC.
Alternative signal timing plans may be used for the following purposes:

- Diversion along preplanned routes in the event of a freeway incident.
- Support of operations to improve throughput in a transportation corridor.

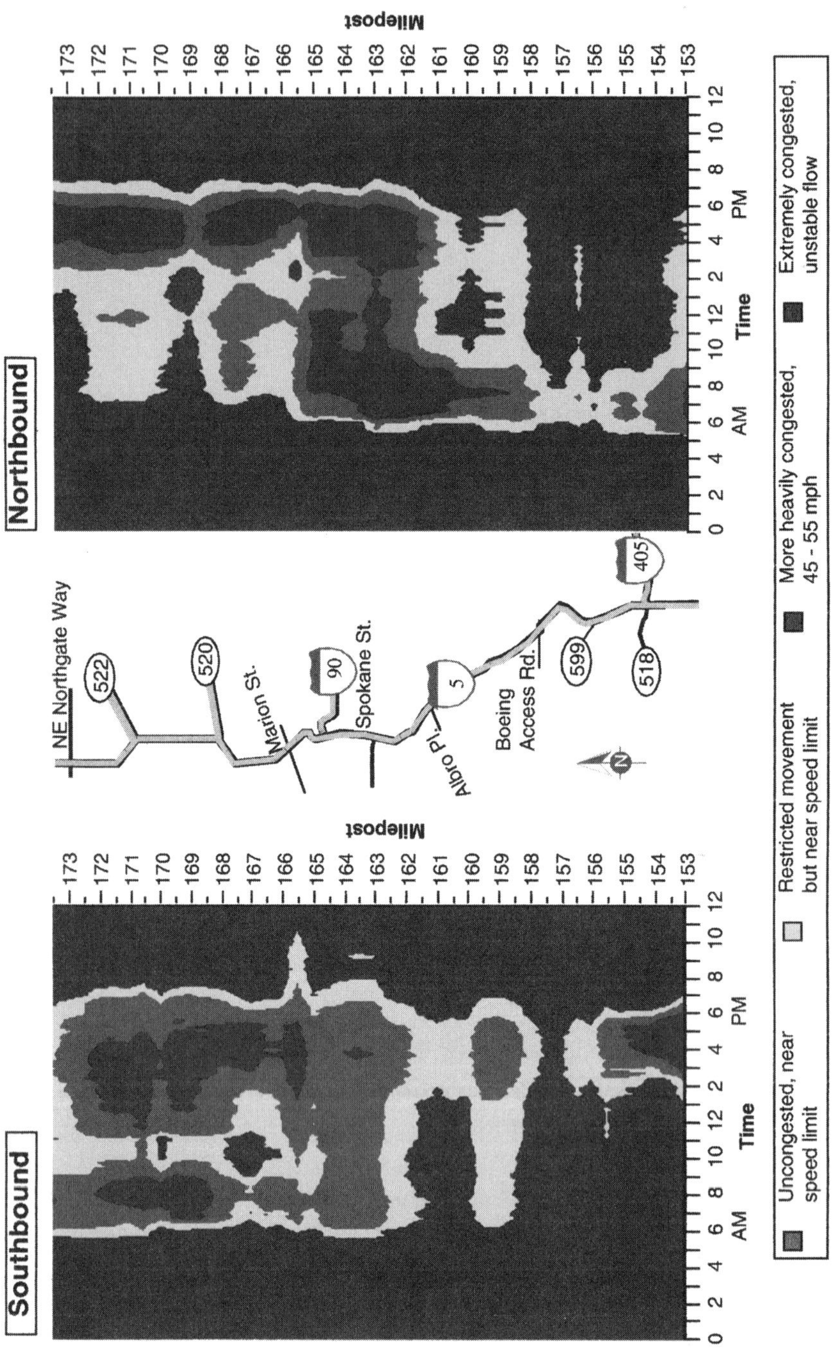

Fig. 9.1 Daily speed profile for roadway section

- Implementation of freeway and surface street traffic signal plans to support special events.
- Implementation of freeway and surface street traffic signal plans to support emergency evacuations.

9.1.6 Provision of Weather Information Related to Roadway Conditions

TMCs provide weather information related to roadway conditions. This information may be provided by systems operated by the TMC, or the TMC may receive this information from a statewide system. Roadway weather information systems (RWIS) utilize measurements made at field stations that may include:

- Air temperature, pressure, and humidity.
- Wind speed and direction.
- Precipitation type and amount.
- Visibility.
- Pavement and subsurface temperature.
- Pavement condition, e.g., dry, damp, wet, ice, snow, residual salt, and water film.

This knowledge is utilized by models that, together with National Weather Service and private weather service data, provide information on visibility and precipitation related conditions. This information warns motorists of hazardous conditions and hence may influence route selection. The information is also used to deploy snowplows and sanders to the most likely incident prone locations.

9.2 Information Flows Among Stakeholders

Four general methods of sharing information for incident management and other purposes are utilized:

- Face-to-face information sharing – This usually takes place when facilities are shared by stakeholders. Improvement in operations results when stakeholder management centers are co-located as co-location provides opportunities for close collaboration in planning operations involving multiple stakeholders. In some cases equipment may be shared among stakeholders. For example, the Regional Traffic Operations Center in the Rochester, N.Y. region (Fig. 9.2) houses the operations centers for NYSDOT freeway operations, the Monroe County traffic signal system (which operates signals in the City of Rochester, Monroe County and interconnected state signals, and the New York State Police. NYSDOT and County CCTV cameras are controlled by a single system.

Fig. 9.2 Regional operations center in Rochester NY

Every camera may be accessed by either agency and displayed either on workstations or on shared large screen displays. Similarly, a single system for controlling CMS is shared by both agencies.

Other resource sharing opportunities include the following [4]:

- Sharing of the same computer aided dispatch (CAD) system by emergency responders.
- Text message interfaces between CAD systems and the freeway management system.
- Sharing communication facilities among stakeholders. This may facilitate display of CCTV at stakeholders facilities or in emergency response vehicles.
- Remote voice – Telephones and land mobile radio are examples of this method.
- Electronic text – Examples include paging, facsimile, email, and text access to traffic incident-related data systems.

– Other media and advanced systems such as video and traffic management systems. Computer-to-computer data transfers may be performed as described in Sect. 8.2 using NTCIP, TCIP OR IEEE 1512 protocols or with other standard data transfer protocols.

This type of data transfer may require reprogramming of legacy systems. The modification cost might be significant, particularly, for metropolitan areas with limited resources. Alternatively, remote workstations offer a relatively inexpensive way to share information. For example, placing a TMC workstation or workstation software in another management center will enable that center to view road network conditions, equipment status, and incident status.

Table 9.1 summarizes the application of these methods for incident management at several locations in the United States.

9.3 Implementation of Information Flows

TMCs are key instruments for implementing the Regional ITS Architecture. This section illustrates how the information flows identified in the Regional Architecture may be implemented by TMCs. The Regional ITS Architecture represents a long-range view of the region's ITS needs. For example, Fig. 9.3 depicts a section of the information flows identified in the regional architecture for the Binghamton, NY metropolitan area.[1]

In contrast, projects that install field equipment or establish or modify TMCs satisfy a subset of these needs. These projects must identify the particular information flows to be incorporated into the project and must define the methodology for accomplishing these information flows. As an example of how this is accomplished, Table 9.2 shows a methodology for implementing information flows related to the TMC described in the Regional ITS Architecture. The specific Regional ITS Architecture information flows are shown in Fig. 9.3 and are included in a project consisting of the design and construction of a TMC and field devices (CMS, HAR transmitters, HAR beacons and traffic detectors.)

9.4 Example of Transportation Management Center in Major Urban Location[2]

Figure 9.4 shows The New York State Department of Transportation's INFORM transportation management center (TMC), which manages traffic on the limited access highways and on a number of surface streets on Long Island. Major facilities

---

[1] The NYSDOT Regional MCO Center in the figure is the New York State Maintenance and Construction Operations Center.

[2] Much of the information in this section was provided by Messrs. Emilio Sosa of the New York State Department of Transportation and Richard Knowlden of Parsons Brinckerhoff, Inc.

Table 9.1 Summary of information sharing methods for incident management by location from NCHRP Report 520: *Sharing Information Between Public Safety and Transportation Agencies for Traffic Incident Management*, p.2

Location	Face-to-face	Remote voice	Electronic text	Other media and advanced systems
Albany, NY	Two co-location sites	Some sharing of public safety radios; some use of commercial wireless service "talk groups"	Shared CAD system	Roadway data, images, and video shared remotely
Austin, TX	Co-location site ready to open	Service patrols equipped with local police radios	CAD data to be shared remotely	CCTV control shared with local police
Cincinnati, OH	Transportation center hosts regional incident management team operations	Some sharing of public safety radios; some use of commercial wireless service "talk groups"	Shared CAD under development	CCTV and other traveler information are shared with public
Minneapolis, MN	Multiple co-location sites	Shared radio system; some use of commercial wireless service "talk groups"	Shared CAD data	CCTV and other traffic management systems are shared
Phoenix, AZ	–	Service patrols equipped with state police and DOT radios	DOT data workstations provided to local public safety agencies	CCTV shared with local fire department
Salt Lake City, UT	Co-location site	Shared radio system	Shared CAD data	CCTV and other traffic management systems are shared
San Antonio, TX	Co-location site	Service patrols equipped with local police radios; shared radio system to be deployed	Shared CAD data	CCTV and other traffic management systems are shared
San Diego, CA	Co-location site	Service patrols equipped with local police radios	Shared CAD data	CAD data are posted on traveler information website
Seattle, WA	–	Service patrols equipped with state patrol radios; center–center intercom system	Shared CAD data	Control of CCTV is shared with state patrol

All locations use standard telephones and facsimile machines for information sharing
CAD computer aided dispatching; *CCTV* closed circuit television; *DOT* department of transportation

Binghamton Regional ITS Architecture
Information Flows

SourceElement	DestinationElement	FlowName
NYSDOT REGIONAL TMC	NYSDOT REGIONAL TMC Personnel	ISP operating parameters
		traffic operator data
	NYSDOT REGIONAL TMC Roadside Equipment	freeway control data
		roadway information system data
		signal control data
		traffic sensor control
		video surveillance control
	BROOME COUNTY Emergency Management Center	archive coordination
		emergency traffic control response
		incident information
		incident information request
		resource deployment status
		road network conditions
	BROOME COUNTY Transit Center	request transit information
		road network conditions
		selected routes
		traffic control priority status
		transit information request
	BROOME COUNTY Transit Center Kiosks	broadcast information
		traveler information
		trip plan
	NYSDOT REGIONAL MCO Center	archive coordination
		field equipment status
		incident information
		maint and constr resource request
		road network conditions
		road network probe information
		work plan feedback
	NYSDOT REGIONAL MCO Field Devices	environmental sensors control
		freeway control data
		roadway information system data
		signal control data
		traffic sensor control
		video surveillance control
	Media	road network conditions
		traveler information for media
	Individual Traveler Personal Computing Devices	broadcast information
		traveler information
		trip plan
NYSDOT REGIONAL TMC Personnel	NYSDOT REGIONAL TMC	ISP operating parameter updates
		traffic operator inputs
NYSDOT REGIONAL TMC Roadside Equipment	NYSDOT REGIONAL TMC	freeway control status
		request for right-of-way
		roadway information system status
		signal control status
		traffic flow
		traffic images
	NYSDOT REGIONAL MCO Center	environmental conditions data
		field device status
		infrastructure monitoring sensor data
		roadway information system status
		traffic images
	NYSDOT REGIONAL MCO Field Devices	roadway equipment coordination
	NYSDOT REGIONAL MCO Vehicles	infrastructure monitoring sensor data
		roadway information system status

Fig. 9.3 Portion of Information Flows for Binghamton, NY. Metropolitan Area Regional ITS Architecture

managed include the Long Island Expressway, Northern State Parkway, and Southern State Parkway. The TMC operates 24 hours a day, 365 days a year. A major function of the TMC is to assist emergency responders in the management and clearance of incidents. Figure 9.5 shows a commercial vehicle that illegally entered a

Table 9.2 Implementation of information links

Source element	Destination element	Flow description	Implementation methodology
NYSDOT Regional TMC	NYSDOT Regional TMC Roadside Equipment	Roadway information system data	CMS NTCIP protocol for controls and message delivery
			HAR voice audio file standard
			HAR beacon control signal
		Signal control data	NYSDOT standard signal system
		Traffic sensor control	NTCIP protocol for data request
		Video surveillance control	NTCIP protocol for camera controls
	Broome County Emergency Management Center	Incident information	NYSDOT TMC workstation, telephone
		Road network conditions	NYSDOT TMC workstation, telephone
	NYSDOT Regional MCO Center	Field equipment status	NYSDOT TMC workstation
		Road network conditions	NYSDOT TMC workstation
		Work plan feedback	Telephone, email
NYSDOT Regional TMC Roadside Equipment	NYSDOT Regional TMC	Freeway control status	CMS Status – NTCIP protocol
			HAR Status – proprietary protocol
			HAR beacon status – simple switch indication
		Traffic flow	Detector data – NTCIP protocol
		Traffic images	CCTV camera images – MPEG standard

Long Island parkway and became trapped under one of the parkway's low overpasses.

Figure 9.6 displays the facility's control room. Services performed by the INFORM TMC include:

- Notice of traffic conditions on 185 changeable message signs These CMS provide indications of both recurrent and non-recurrent congestion events. Messages for over 5,000 incidents per year are provided.
- Information dissemination to the media, the NYSDOT 511 system and to other transportation agencies.
- Monitoring of roadway conditions via 192 closed circuit television cameras. A number of these cameras are located on surface streets and near special event venues. Public access to the camera displays is available through web sites.

Fig. 9.4 INFORM Traffic Management Center. Source: Parsons Brinckerhoff, Inc.

Fig. 9.5 Incident on Long Island parkway. Source: Parsons Brinckerhoff, Inc.

- Collection of information from over 3,000 traffic detectors including inductive loop detectors, video detectors, and acoustic detectors. This information assists in incident management and development of motorist information. The system also operates nine sites providing weather information.
- Operation of 91 entry ramp meters. Figure 9.7 shows a typical ramp meter installation. These ramp meters are generally operated in a non-restrictive fashion (Sect. 7.1.)

Fig. 9.6 INFORM control room. Source: Parsons Brinckerhoff, Inc.

Fig. 9.7 INFORM ramp meter. Source: Parsons Brinckerhoff, Inc.

- Twelve travel time signs at key locations on the Northern State Parkway.
- Coordination of 570 traffic signals on surface streets that support the limited access highway network. In the event of incidents or construction on the limited access highways, emergency signal timing plans facilitate diversion to these facilities.
- Motorist assistance during morning and evening peak travel periods. The motorist assistance vehicles perform over 27,000 stops and provide over 7,600 assists annually.

References

1. Transportation management center concepts of operation: implementation guide (1999) Federal Highway Administration, Washington, DC
2. Ishimaru JM et al (1999) Central Puget Sound freeway network usage and performance, 1999 update, vol 1. Report WA-RD 493.1. Washington State Transportation Center Seattle, WA
3. Beaulieu M et al (2007) A guide to documenting VISSIM-based microscopic traffic simulation models. Report WA-RD 678.1. Washington State Transportation Center Seattle, WA
4. Brooke K et al (2004) Sharing information between public safety and transportation agencies for traffic incident management. NCHRP report 520, Transportation Research Board, Washington, DC

Chapter 10
Evaluation of System Design and Operation

Abstract Prospective evaluations (evaluations prior to system installation) and retrospective evaluations (evaluations after system operation commences) are necessary to ensure that proposed and implemented systems satisfy the stakeholders' requirements upon which they are based. To be effective, these evaluations require methods for performing annualized benefit and cost analyses. Selection of parameters for assessing benefits, and identification of key parameters associated with benefits estimation are important for these evaluations. These topics, along with planning requirements for project evaluations after the commencement of the operation and the impact of evaluation requirements on system design, are discussed in this chapter.

10.1 Evaluation of Design Alternatives and Project Feasibility

Prospective evaluations are used to evaluate design alternatives and project feasibility during the project's planning and design phases. These evaluations usually utilize benefit and cost analysis as well as consideration of qualitative factors that are important but do not lend themselves to this type of analysis.

10.1.1 Benefit and Cost Analysis

Benefit and cost analysis has been the principal tool that is traditionally used during the initial project planning and later design phases for highway related projects. As used in prospective assessments, benefit and cost analysis assists with the evaluation of candidate design alternatives and in establishing the feasibility of the project (possibly in relation to other projects that compete for resources). It is also used to establish priorities among projects competing for available funding. As applied to retrospective assessments, benefit and cost analysis assists in evaluating the

R. Gordon, *Intelligent Freeway Transportation Systems: Functional Design*,
DOI 10.1007/978-1-4419-0733-2_10,

improvements to system operations, establishing lessons learned, providing inputs to reports to the public, and improving and selecting future designs and projects. The following sections describe the application of benefit and cost analysis to freeway ITS projects.

10.1.1.1 Methodology for Cost Estimation

A commonly used methodology is an annual cost comparison [1]. This entails converting capital costs to an equivalent annual cost and adding the annual maintenance and operating expenses.

A useful life for the project must be established. For ITS projects it is often considered to be in the range of 20–25 years. Although much equipment will likely be replaced before this period, these replacements may be generally considered as a maintenance expense for the purpose of this analysis.

Several techniques to annualize capital cost are available. The frequently applied capital recovery with a rate-of-return method is described below. This approach converts the cost of a design alternative into an equivalent uniform series of annual costs. The salvage value after the project's useful life is generally considered as zero. The conversion is provided by the following equation:

$$R = \mathrm{PC} \bullet \mathrm{crf} \tag{10.1}$$

Where R is the annualized cost with interest, PC the capital equipment cost, and crf is the capital recovery factor.

The capital recovery factor is a function of the interest rate (i) and the project useful life in years (n). It is provided in tables in engineering economics texts and its expression is shown below:

$$\mathrm{crf} = \mathrm{i} \bullet (1 + \mathrm{i})^{\mathrm{n}} / ((1 + \mathrm{i})^{\mathrm{n}} - 1) \tag{10.2}$$

The annualized life cycle cost is the sum of (10.1) and the annual maintenance and operating expenses.

10.1.1.2 Implementing Benefit and Cost Analysis

Benefit estimations may be performed by ITS evaluation programs (Sect. 3.4.1), by use of the DOT ITS data base [1] or by extension from previous experience. Figure 10.1 shows the results of a benefits analysis for a number of functional design alternatives for a project. The benefits were obtained using the Design ITS model described in Sect. 3.4.1.1.

Although the benefit-to-cost ratio is a commonly used measure of the relative value of a project or of alternative designs, this ratio must be applied in conjunction with other factors, including the ability to achieve project objectives. It is helpful to

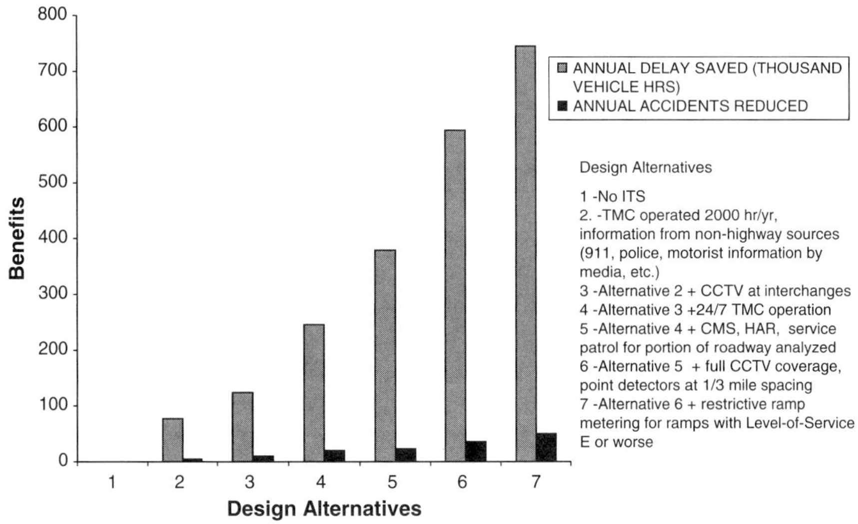

Fig. 10.1 Benefits analysis

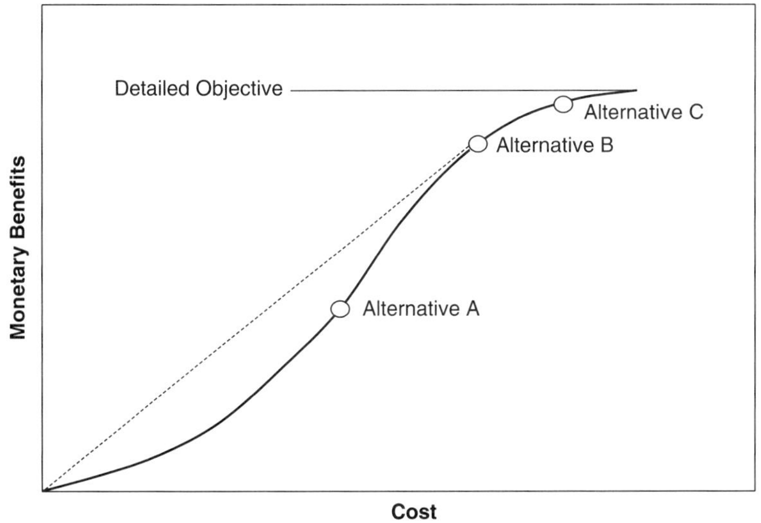

Fig. 10.2 Monetary benefits and costs for design alternatives

depict the benefits and costs in a more general way, such as is conceptually shown in Fig. 10.2. The annualized benefits and life cycle costs are shown for alternatives A, B and C. Each of these alternatives has a benefit-to-cost ratio considerably in excess of 1.0. While Alternative B has the highest benefit-to-cost ratio (depicted in the figure as the slope of the dashed line), Alternative C more closely approaches the detailed objective.

Selection between Alternatives B and C may be assisted by performing a of marginal benefit to cost analysis as discussed in Sect. 3.2.1. The example shown in that section may be generalized to include all monetary benefits.

10.1.1.3 Selection of Parameters for Estimation of Benefits

When evaluation programs such as those described in Sect. 3.4.1 are used for benefits estimation, the estimate may be quite sensitive to the parameters assumed. The following example illustrates the sensitivity to link volumes. The volume-to capacity-ratio (q/C) for the highest volume case (considered as the baseline) is shown in Table 10.1.

The example uses the ITS treatments shown in Fig. 2.3. The baseline case describes a high volume-to-capacity situation for all links except for the short link. Figure 10.3, developed by using the Design ITS model, shows the strong sensitivity of delay reduction to the values of mainline volumes assumed when q/C exceeds 0.9. In order to improve the credibility of the analysis, it may be constructive to conduct the analysis using a range of volume inputs.

To obtain the present worth of future year benefits, each future year benefit must be multiplied by the single payment present worth factor (sppwf) [1]. This is given by (10.3)

$$\text{sppwf} = 1/(1 + i)^{n-0.5} \tag{10.3}$$

where i is the interest rate and n is the year for which the sppwf is computed. Equation 10.3 uses the term n–0.5 instead of the more commonly used n in order to represent the average annual benefit more closely. Although it is more accurate to conduct the analysis by employing each year's volumes to compute the benefits for that year, a commonly applied approximation is to select a single appropriate future year and utilize the volumes for that year.

Because the benefits for the more distant years are more lightly weighted, the appropriate set of volumes to use for the benefits evaluation should be more heavily biased toward the early years for the project. As an aid to selecting the most appropriate year's volumes to employ, Table 10.2 identifies the year in which 50% of the

Table 10.1 Parameters for baseline case

Link	Link length (miles)	Volume to capacity ratio (q/C)
1	0.88	0.65
2	3.12	0.94
3	3.60	0.94
4	1.31	0.96
5	1.11	0.93

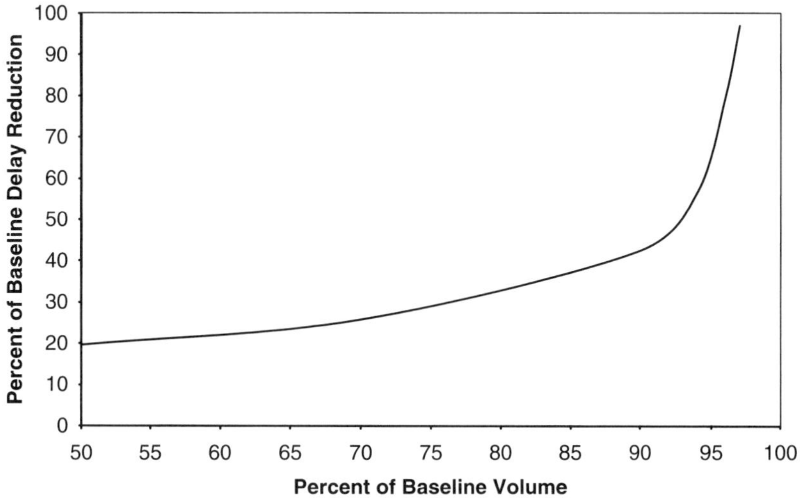

Fig. 10.3 Delay reduction versus assumed volume

Table 10.2 Year in which 50% of total project benefits are achieved

Interest rate (%)	Year after project initiation
4	8.1
6	7.2
8	6.5

total life cycle benefits are accumulated after operation commences. Table 10.2 is based on a 20-year project life.

10.1.2 Alternatives Evaluation and Project Feasibility

The development of the Transportation Improvement Plan (TIP)[1] requires the consideration of candidate ITS projects relative to other candidate transportation improvement projects. Similarly, the development of the Regional ITS Architecture and the selection of actual projects to implement the architecture requires consideration of alternative projects. Stakeholders and decision makers are key contributors to the project selection process. The benefit and cost considerations described in Sect. 10.1.1 usually play a significant role in the process, but there are other considerations as well. Section 3.2.1 identifies the following considerations that are common to ITS projects described in this book:

[1] To qualify for federal aid funding, the project must be included in the Transportation Improvement Plan prepared by the metropolitan planning organization.

- Economic benefits. For evaluation purposes, delay reduction, safety, and fuel consumption are considered in this category.
- Environmental benefits (emissions).
- Mobility. This represents the ease with which people and goods move [3]. Measures may include delay and variation in delay [4].
- Public satisfaction with ITS treatments.

Section 3.2.2 describes multi-attribute utility analysis as an evaluation tool that facilitates the participation of stakeholders and decision makers in the selection of projects and functional alternative project designs.

10.2 Project Evaluation

Project evaluations are retrospective evaluations, i.e., they are implemented after the design of the project. Portions of the evaluation may be conducted prior to or during project installation. The reasons for performing project evaluations include:

- Improvement of operations. Lessons learned during the evaluation may provide the basis for improved operations.
- Reports to the public. Public support for ITS is crucial for its continued success.
- Improvement of future designs.
- Assistance in the selection of future projects [3].

ITS project operations are generally envisioned to continue for an indefinite time period. The evaluation process should therefore be considered as a continuous or at least periodic function, in order to keep the project in tune with evolving objectives, requirements, and technologies. This is schematically illustrated in Fig. 10.4 by the addition of an evaluation feedback function to the simplified Vee diagram of Fig. 2.1.

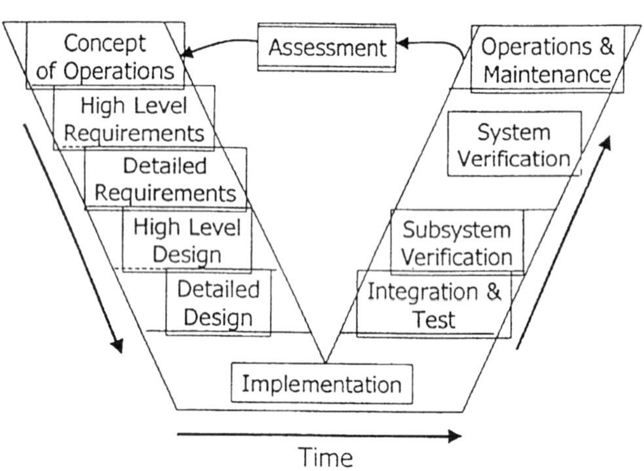

Fig. 10.4 Vee diagram with evaluation feedback

Evaluation planning should be considered during the system design process. If "before" conditions are to be measured in the field, the plan may require data collection prior to project installation. Primary methods of data collection include: field observation, automatic data collection devices, simulations, and surveys [4]. It may be useful to use traffic system detectors as a means of collecting volume data, speed data, and day-to-day variation in speed. After project operation commences, some period of time is required for traffic conditions to stabilize. This occurs when the benefits do not change significantly over short periods of time. This time period is usually considerably less than 1 year [4].

Evaluation may be facilitated by including data mining software into the system design. The archived data user service (ADUS) has been incorporated into the National ITS Architecture [5] and guidelines for its use have been developed [6]. For example, the PORTAL ADUS system, developed by Portland State University is extensively used for evaluation in Portland, Oregon [7]. It archives and analyzes inductive loop detector data to create a variety of reports.

References

1. Maccubin RP et al (2003) Intelligent transportation systems benefits and costs: 2003 update, Report FHWA-OP-03-075. Mitretek Systems, Inc., Washington, DC
2. Taylor GA (2005) Managerial and engineering economy, 2nd edn. D. Van Nostrand Company, New York, NY
3. Highway mobility operations manual (2005) Oregon Department of Transportation Salem, OR
4. Development of a project evaluation methodology framework for Canadian Intelligent Transportation Systems (2007) Delcan
5. ITS data archiving: five-year program Description (2000) U.S. Department of Tranportation, Washington, DC
6. Turner S, Guidelines for developing ITS data archiving systems (2001) Report 2127-3. Texas Department of Transportation, Texas Transportation Institute, US DOT
7. Bertini RL et al (2005) Experience implementing a user service for archived intelligent transportation systems data. Transportation Research Record 1817:90–99

Chapter 11
Compact Disk

Abstract This chapter provides short descriptions of the files on the compact disk that accompanies this book. The files implement models that are described in more detail in earlier chapters.

11.1 Introduction

In order to facilitate the reader's use of several of the models described in the text, several worksheets are provided in the form of Microsoft® Excel workbook files and a Microsoft® Word file on the compact disk that accompanies this book. The contents of these files are described in the following sections. In order to facilitate modifications or adaptations that the reader deems appropriate, the Excel workbook files are not protected and are populated with data to illustrate the type of data needed. Readers must substitute their own data to obtain results specific to their applications.

11.2 System Delay per Incident

Since incidents may occur under a variety of traffic conditions, the model offers a methodology to apportion the appropriate fraction of daily volume to the capacity conditions that apply and to then compute the resulting delay. Two worksheets are used to estimate the delay resulting from an incident on a roadway section with three lanes in one direction.

As described in Sect. 4.4.1 and illustrated in Fig. 4.12, the first worksheet (CD Excel file *Cohort Factors 3 Lane*) Titled "COMPUTATION OF COHORT FRACTIONS (THREE LANE ROADWAY)" generates fractions for the relative

R. Gordon, *Intelligent Freeway Transportation Systems: Functional Design*,
DOI 10.1007/978-1-4419-0733-2_11, © Springer Science+Business Media, LLC 2009

occurrence of incidents for groups of volume-to-capacity ratios (cohorts). This file is applicable where the roadway section under analysis consists of three lanes in one direction The user must supply appropriate hourly volumes in columns 2 and 3, appropriate roadway section titles and an appropriate capacity. The data from the file are utilized as input to the CD Excel file *Average delay resulting from incident.* Figure 4.13 illustrates the worksheet and Sect. 4.4.2 describes the inputs to the file.

11.3 Relative Effectiveness of CCTV Coverage

Section 4.5.1.2 describes the considerations for providing CCTV coverage. Appendix B presents a model for the effectiveness of this coverage. The measure, RTV represents the probable capability of CCTV to observe an incident on a section of roadway. Figure 4.15 shows an application of this model to a particular case. The Excel file *RTV* on the CD provides one data point for this figure. The shaded cells require data input by the analyst.

11.4 Incident Management Effectiveness Potential

A model for assessing the effectiveness of ITS in enhancing the management of the response to incidents is described in Sect. 4.6.3.2. Figure C.1 in Appendix C illustrates an application of the model to evaluate candidate system design alternatives. Computations for this figure are shown in Fig. C.2. The CD Excel file *Inc mgt effectiveness potential* was used to compute the data for the 3 camera, 5 detector alternative for Fig. C.1. The value of RTV was obtained from Figure 4.15

11.5 Delay Reduced on Freeway due to Queue Reduction Resulting from Diversion

Section 5.1.5 describes a model that provides approximate values for the reduction of freeway delay resulting from diversion. Figure 5.6 shows the results of the Excel worksheet file *Delay improvement on freeway.* The shaded cells in the workbook represent required data entries. The nonshaded cells perform the worksheet calculations.

11.6 Probability that the Motorist Encounters CMS Prior to Incident (P34)

Section 5.2.2.2 illustrates the computation of the parameter P34 when origin–destination data is not available. This parameter represents the probability that the motorist encounters a CMS prior to the section containing the incident, thus providing the capability for diversion. The worksheet shown in Fig. 5.12 was computed by the Excel file *Computation of P34.*

11.7 Queue Storage Requirement for Ramp Meter

Section 7.4.5.1 describes Caltrans' approach to estimating the storage space required for metered ramps. Figure 7.15 illustrates a typical computation chart [1]. The reference also provides a chart that is not populated with data. For convenient use, that chart is reproduced in the Word file *Ramp storage empty computation chart.*

Reference

1. Ramp meter design manual (2000) Traffic operations program, California Department of Transportation, Sacramento, CA

List of Symbols, Abbreviations and Acronyms

Acronym	Definition
AADT	Annual average daily traffic
ACCR	Accident rate
ACR	Accident rate for section
ADD	Average vehicle delay (diversion)
ADND	Average vehicle delay (no diversion)
ADUS	Archived Data Use Service
AID	Automatic incident detection
ALINEA	A local rump metering algorithm
AM	Amplitude modulation
AR	Accidents reduced
ARI	Accident rate in interchange area
ARNI	Accident rate in non-interchange area
ARTEMIS	Advanced Regional Traffic Interactive Management and Information System
ATSMR	Average time per mile per vehicle saved by metering
$B1$	Timely detection probability for a range of scenarios for different detector spacing
B_J	Total mainline traffic in section J
C	Capacity
CAD	Computer-aided dispatch
CALTRANS	California Department of Transportation
CCTV	Closed circuit television
CD	Capacity deficit
CDMA	Code division multiple access
CDS	Total corridor delay reduction
CFR	Code of Federal Regulations
CMS	Changeable message sign
CO	Carbon monoxide

Acronym	Definition
CONOPS	Concept of operations
crf	Capital recovery factor
CS	Capacity along arterial with signal
CVF	Commercial vehicle fraction
D_I	Delay from the start of the incident up to incident clearance
D_Q	Delay from incident clearance to queue dissipation
D_T	Total delay
DAR	Additional delay incurred by pre-diversion traffic on diversion route
DC	Delay from TC to T
DD	Delay prior to TC
DF	Maximum diversion fraction for no major arterial congestion
DIF	Delay for non-diverted freeway traffic
DOT	Department of Transportation
DQC	Delay after incident clearance
DTA	Dynamic traffic assignment
EMFITS	Evaluation Model for Freeway ITS Scoping
EMI	Electromagnetic interference
ER	Emission rate
E511	Emergency telephone response service provided by PSAP
FHWA	Federal Highway Administration
FM	Frequency modulation
FRM	Fraction of roadway segment influenced by metering
FRR	Fraction of ramps in roadway segment that contain ramp meters
G	Gini coefficient
GHz	Gigahertz
GPS	Global Positioning System
GS	Green split along arterial
GSM	Global System for Mobile Communications
H	Potential ability for ITS to provide incident management support
HAR	Highway advisory radio
HC	Hydrocarbons
HOT	High occupancy toll
HOV	High occupancy vehicle
i	Interest rate

Acronym	Definition
IDAS	ITS Deployment Analysis System
IDI	Duration of incidents and accidents in interchange areas
IDN	Duration of incidents and accidents in non-interchange areas
IDV	Time saved by non-diverting vehicles
IEEE	Institute of Electrical and Electronics Engineers
IIMS	Integrated Incident Management System
INCOSE	International Council on Systems Engineering
INFORM	Information for Motorists
ITS	Intelligent Transportation Systems
k	Density or concentration
K	Parameter
KD	Constant for utility calculation
KS	Satisfaction rating
LD	Length of detector sensing area
LOS	Level of service
LS	Section length
LV	Length of vehicle
L1	Distance in the section in the vicinity of the upstream interchange for the section midpoint of the interchange that encompasses most of the accidents
L2	Distance in the section in the vicinity of the downstream interchange
MCO	Maintenance and Construction Operations Center
M_D	Mainline AADT for section with upstream CMS closest to section being analyzed
MDF	Maximum divertible freeway flow without major arterial congestion
MF	Maximum flow without major arterial congestion
M(*J*)	Mainline AADT for section *J*
MPEG	Moving Picture Experts Group
MTF	Fraction of planned metering period that freeway is at level of service E or worse
MTTSV	Average delay reduced per assisted vehicle
MU	Multivariate utility value
MUTCD	Manual of Uniform Traffic Control Devices
MV	Utility for message type
MVM	Million vehicle miles
MVMPY	Motor vehicle miles per year
n	Project useful life

Acronym	Definition
NAIR	Non-accident incident rate
ND	Total number of diverted vehicles until incident clearance
NFA	Normal background traffic
NMV	Utility for other than message type
NND	Total number of vehicles served until incident clearance
NOX	Oxides of nitrogen
NT	Number of travelers
NTCIP	National Transportation Communications for ITS Protocol
NTOC	National Transportation Operations Coalition
NYSDOT	New York State Department of Transportation
OREMS	Oak Ridge Emergency Evacuation System
OSI	Open Systems Interconnection (model)
P	Probability of functional capability
P3	Probability motorist receives and understands message from all media
P4	Probability that qualifying motorist receiving and understanding the message diverts
P10	Probability incident is managed if TMC is staffed
P21	Probability TMC staffed when incidents occur
P34	Probability that motorist encounters CMS prior to reaching diversion location
P35	Probability that motorist will read sand understand CMS message
PC	Capital equipment cost
PD	Propensity to divert
PHV	Peak hour volume
PORTAL	Portland Oregon Regional Transportation Archive Listing
PSAP	Public service access point
P(x)	Cumulative probability distribution
q	Volume
q_1	Volume at incident
q_2	Volume entering incident
q_3	Volume when incident is present
qA	Normal arterial volume
q/C	Volume-to-capacity ratio
QUAL	Relative quality of information
R	Annualized cost with interest

Acronym	Definition
RC	One direction arterial capacity
RE	Residual capacity
RI	Project class
R(*j*)	Meter rate after time *j*
R(*J*)	Residual volume for section *J* after exit ramps
RTV	Relative coverage of incidents
S_j	Speed of vehicle *j*
S_1	Volume at incident clearance (roadway capacity)
S_2	Volume entering incident location (demand volume)
S_3	Volume when incident is present (restricted capacity resulting from incident)
SCRITS	Screening Analysis for ITS
sppwf	Single payment present worth factor
SWARM	Systemwide Adaptive Ramp Metering Algorithm
t_j	Occupancy period sensed by detector for vehicle *j*
T	Time from start of incident to incident clearance
TA	Total number of accidents in section
TC	Time from incident occurrence until diversion is implemented
TCIP	Transit Communications Interface Profiles
T_D	Time for queue to dissipate after incident starts
TDD	Time period after incident clearance until queue clears under diversion conditions
TELCO	Local franchised wireline telephone company
TIP	Transportation Improvement Plan
TMC	Transportation (or traffic) management center
TNA	Test for presence of non-interchange area in section
TO	Metering hours per year
TOND	Improvement in delay on diversion route
TS	Delay reduced
TSI	Delay reduced per incident
TSVD	Time saved by diverting vehicles
TVI	Fraction of roadway visible to CCTV at interchanges
TVN	Fraction of roadway visible to CCTV away from interchanges
T1	A data channel that runs at 1.544 Mbits/s data rate
U	Utility value or difference in utility values
u	Mean-space-speed
u_T	Mean-time-speed
V	Volume

Acronym	Definition
VA	Mainline volume upstream of queue formed by incident
VAI	Entry volume downstream of CMS and upstream of incident
VD	Diversion volume
VII	Vehicle Infrastructure Integration
W_{AB}	Velocity of shock wave propagation. A is downstream boundary, B is upstream boundary
x	Queue discharge volume
X(J)	Sum of exit ramp AADT for section J
Y	Fraction of incident management operation represented by timely assistance to emergency responder function
511	Telephone number for public service answering point
α	Shape parameter
β	Scale parameter
θ	Occupancy
θ_I	Filter input data
θ_M	Mainline detector occupancy
θ_O	Filter output data
θ_S	Occupancy setpoint

Appendix A: Sources of Additional Information on Evaluation Models

1. Design ITS

This model was developed by the author. Additional information on the model and how to acquire it can be found on the model's website: http://designints.com/

2. Evaluation Model for Freeway ITS Scoping (EMFITS)

EMFITS was developed by Dunn Engineering Associates for the New York State Department of Transportation Project Development Manual (PDM). The documentation appears as Supporting Documentation in the PDM.

3. ITS Deployment Analysis System (IDAS)

IDAS was developed by Cambridge Systematics for the Federal Highway Administration. Information about the program and how to acquire it is available at the FHWA website: http://ops.fhwa.dot.gov/trafficanalysistools/idas.htm or at the Cambridge Systematics website: http://www.camsys.com/tp_planpro_idas.htm

4. Screening Analysis for ITS (SCRITS)

SCRITS was developed by Science Applications International Corporation (SAIC) for the Federal Highway Administration. The program and its documentation may be downloaded from the following website: http://www.fhwa.dot.gov/steam/scrits.htm

Appendix B: Relative Effectiveness of CCTV Coverage

Most of the accidents on urban and suburban freeways occur in the vicinity of interchanges. These accidents often take longer to clear than non-accident-related incidents. Thus, ITS designs that do not provide for complete or near-complete CCTV coverage usually place most of the cameras in the vicinity of the interchanges. The incident-hours covered by these cameras provide a larger fraction than would be experienced if the accidents were distributed evenly along the roadway (non-accident-related incidents are assumed to be evenly distributed along the roadway). This appendix describes a measure (RTV) for evaluating the relative coverage of incidents by CCTV, and is based on the model included in the Design ITS model. RTV approximately represents the fraction of incident periods in the section that is observable by the CCTV cameras.

Figure B.1 provides the physical layout for the RTV computation. The section shown in the figure represents one direction of the roadway between two interchanges. For convenience, half the distance within the interchange is attributed to the section (the other half is assigned to the adjacent section).

The following parameters are used for the computation of RTV:

ACR Accident rate for section
ARI Accident rate in interchange area
ARNI Accident rate in non-interchange area
IDI Duration of incidents and accidents in interchange areas
IDN Duration of incidents and accidents in non-interchange areas
LS Section length
L1 Distance in the section in the vicinity of the upstream interchange for the section midpoint of the interchange that encompasses most of the accidents
L2 Distance in the section in the vicinity of the downstream interchange; the values L1 and L2 represent half the distance centered at the midpoint of the interchange that encompasses the highest accident rate portion of the section
NAIR Non-accident incident rate
TVI Fraction of roadway visible to CCTV at interchanges

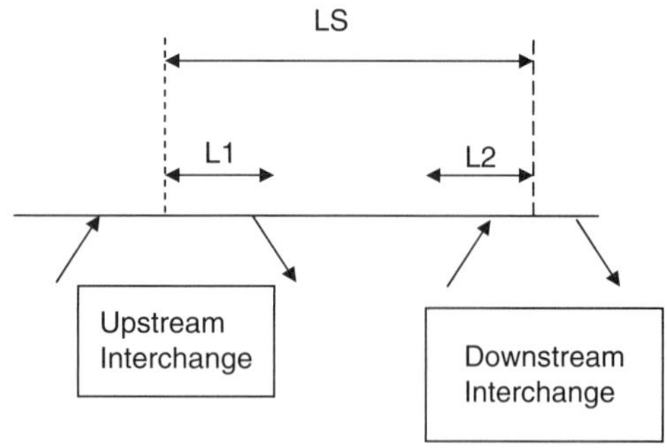

Fig. B.1 Physical relationship for RTV equation

TVN Fraction of roadway visible to CCTV away from interchanges
TNA Test for presence of non-interchange area in section
WE Ratio of accident rate in interchange area to accident rate in non-interchange area

Accident data may be available in the following ways:

- Accident rate summary for the section.
- Detailed accident rate data may be available by milepost (usually 0.1 mile intervals) or by geodetic reference.

Accident rate data for the section (identified above as ACR) may be disaggregated into the high accident rate portion near the interchanges (ARI) and the lower accident portion away from the interchanges (ARNI). The model expresses the approximate relationship.

$$\text{ARI} = \text{WE} \bullet \text{ARNI} \tag{B.1}$$

An estimated value for WE is required. The total number of annual accidents (TA) in the section may be expressed as

$$\text{TA} = \text{ACR} \bullet \text{LS} = (\text{L1} + \text{L2}) \bullet \text{ARI} + (\text{LS} - \text{L1} - \text{L2}) \bullet \text{ARNI} \tag{B.2}$$

Substituting (B.1) into (B.2) yields

$$\text{TA} = \text{AR} \bullet \text{LS} = (\text{L1} + \text{L2}) \bullet \text{WE} \bullet \text{ARNI} + (\text{LS} - \text{L1} - \text{L2}) \bullet \text{ARNI} \tag{B.3}$$

Simplifying (B.3) yields

$$\text{ARNI/ACR} = \text{LS}/(\text{LS} + (\text{L1} + \text{L2}) \bullet (\text{WE} - 1)) \tag{B.4}$$

In some cases detailed accident data at tenth of mile intervals is available, thus providing WE directly from (B.1).

The computations for the relative coverage of incidents by CCTV are as follows:

$$\text{IDI} = (\text{ARNI} \bullet \text{W} \bullet \text{E} + \text{NAIR}) \bullet (\text{L1} + \text{L2}) \tag{B.5}$$

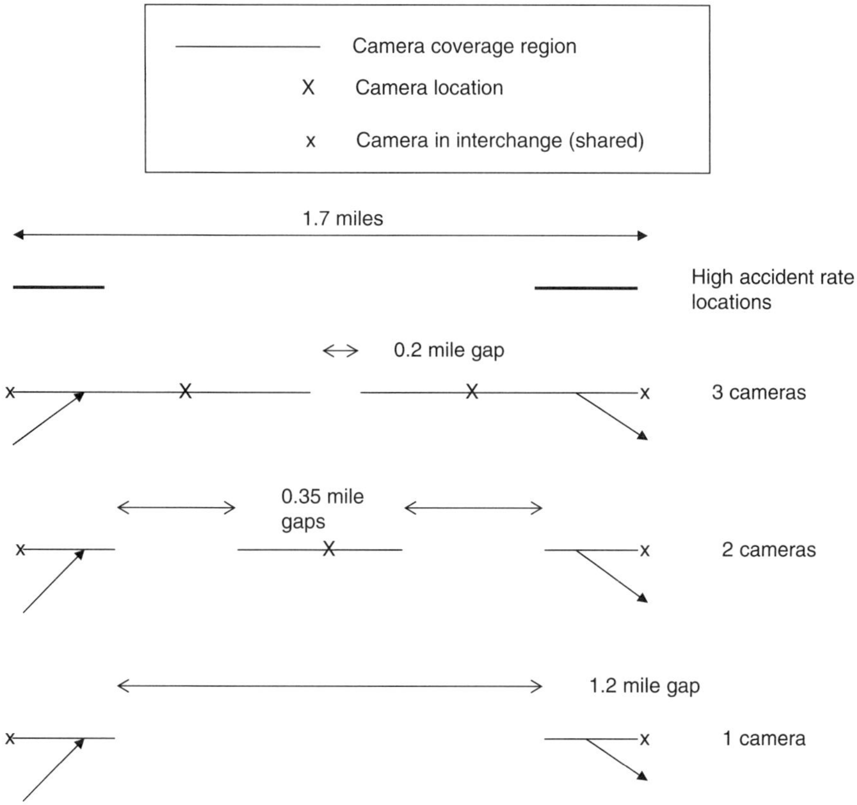

Fig. B.2 CCTV coverage for Camera Development Alternatives

$$IDN + (LS - (L1 + L2)) \bullet (ARNI + NAIR) \tag{B.6}$$

$$RTV + (TVI \bullet IDI + TVN \bullet IDN)/(IDI + IDN) \tag{B.7}$$

Figure B.2 illustrates a simplified case study for evaluating RTV. Here each CCTV camera is assumed to cover one half mile of roadway. A distance of one half mile centered at the midpoint of the interchange is also assumed to represent the accident clustering region. The boundaries for the section under evaluation extend from the midpoint of the upstream interchange to the midpoint of the downstream interchange. Three alternate deployments providing less than 100% CCTV coverage are shown in Fig. B.2, along with the CCTV camera count. The cameras at the interchanges are each counted as one half of a camera because 50% of their coverage extends into another section. Other parameters for the case study example include the following:

- Non-accident incident rate = 7.03 incidents per mile per year
- Accident rate for section = 2.1 accidents per mile per year

	RELATIVE EFFECTIVENESS OF CCTV COVERAGE		
	Data Entry Required		
Scenario:	Section 1, 3 cameras		
Section Length		LS	1.5
Non-Accident Incident Rate		NAIR	7.03
Accident Rate		ACR	2.1
Extension factor for accident clearance		WE	2.40
Upstream Interchange accident range (miles)		L1	0.25
Downstream interchange accident range (miles)		L2	0.25
Fraction of roadway covered by CCTV at interchanges		TVI	1
Fraction of roadway covered by CCTV away from interchanges		TVN	0.83
Relative incident duration in interchange areas		IDI	5.23
Relative incident duration in non-interchange areas		IDN	8.46
Accident rate in non-interchange area		ARNI	1.43
Fraction of roadway incident periods covered by CCTV		**RTV**	**0.89**
if TMC staffed			

Fig. B.3 Relative effectiveness of CCTV coverage

- Ratio of accidents in interchange area to accidents in non-interchange area = 2.4

The worksheet *RTV* on the enclosed CD and shown in Fig. B.3 was used to compute the values shown in Fig. 4.15. If the milepost based accident data is available, the value WE may be computed from this data (B.1) and entered into the worksheet. If only section summary accident rate data is available, the default value of WE = 2.4 may be retained or another value may be substituted.

Appendix C: Example of Benefits for Incident Management

This appendix provides an example of an alternatives analysis for the computation of the reduction in vehicle delay and accidents for ITS support of incident management.

The traffic parameters in the travel direction for the example are described in Table C.1. The parameters for the ITS treatments are shown in Table C.2. Two alternatives are considered for CCTV camera placements and two alternatives for detector deployment are examined in the example.

Equations (4.10)–(4.14) were applied to the parameters described above for the example for alternatives ac, ad, bc and bd. Figure C.1 shows a plot of the results for the alternatives shown in Table C.2 as well as the alternative of no detectors and no CCTV camera. The *Inc mgt effectiveness potential* worksheet on the enclosed CD (shown in Figure C.2) was used to develop the data for the three camera and five detector alternative design in Figure C.1. See figure Fig. C.2.

Table C.1 Traffic parameters for example

Symbol	Definition	Value	Comments
	Number of lanes	3	
AADT	Average annual daily traffic (vehicles/day)	75,000	
ACCR	Accident rate (accidents per million vehicle miles)	2.1	Based on New York state average for freeway accidents
CS	Capacity (vehicles/h)	6,300	
IR	Capacity reducing incidents in incidents per million vehicle miles	9.01	Based on data for upstate New York metropolitan areas
LS	Section length (miles)	1.7	
MVMPY	Million vehicle miles per year	46.54	Based on MVMPY = AADT · LS · 365/1,000,000
PHV	Peak hour volume (vehicles/h)	6,000	
TSI	Time reduced per incident in vehicle miles for high level of ITS deployment and intensive incident management by the TMC	271.8	Table 4.3 for three lanes and traffic condition level 3
K35	Correction factor for level of service	1.0	See Appendix F for further details

Table C.2 ITS parameters for example

Technology or operation	Alternative	Deployment	Comments
911/PSAP information availability		Included in project	
Police operations		Included in project	
CCTV	a	3 cameras	RTV = 0.89 (example in Appendix A and Fig. 4.16)
	b	2 cameras	RTV = 0.62 (example in Appendix A and Fig. 4.16)
Motorist service patrols		Not included	
Electronic traffic detection	c	5 detector stations	Average spacing = 1.7/5 = 0.34 miles/detector B1 = 0.9 (see Sect. 4.5.1.3)
	d	None	No detectors B1 = 0 (see Sect. 4.5.1.3)
TMC operational support		TMC assists in incident management	P10 = 1.0 K5 = 0.1 (see Table 4.12)
		TMC staffed around the clock	P21 = 1.0

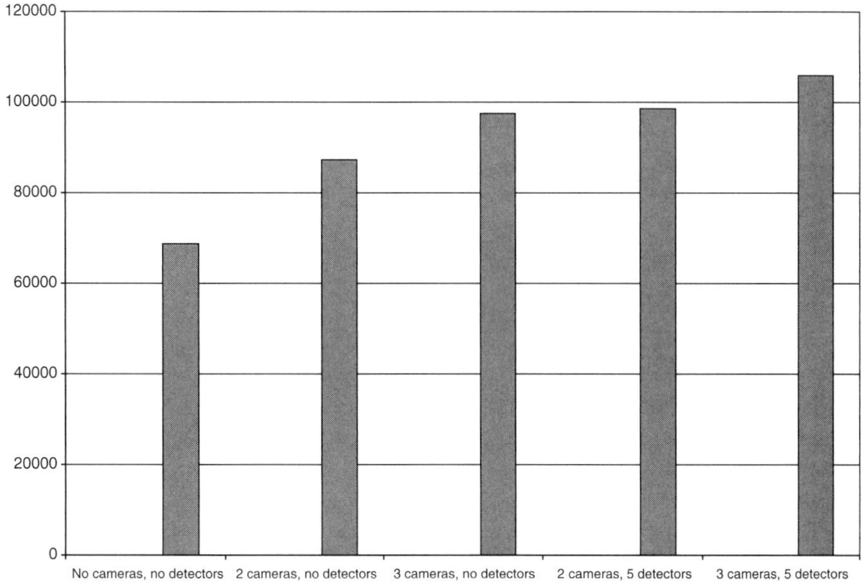

Fig. C.1 Reduction in vehicle hours of delay per year for design alternatives

INCIDENT MANAGEMENT EFFECTIVENESS POTENTIAL						
	Must enter data to obtain results following data entry					
	May be optionally entered or changed by user					
Alterfnative:		3 CCTV cameras in section, 5 detector stations in section, service patrols				
RTV =	0.89					
K40 =	1					
K35 =	1					
B1 =	0.9					
		V_{1g}	V_{2g}	V_{3g}	V_{4g}	V_{5g}
Functions (g)						
1		0.6	0.3	0.9	0.5	0.4
2		0.3	0.6	0.9	0.5	0.2
3		0	0.9	0.8	0.5	0.2
4		0	0.1	0.5	0.2	0.8
	Cumulative Probability for Each Function (H_g)					
Functions	Symbol					
1	H1 =	0.982				
2	H2 =	0.977				
3	H3 =	0.988				
4	H4 =	0.888				
Fraction of benefits obtained by assistance in clearing incident						
Y =	0.8					
Incident Management Effectiveness Potential						
H =	**0.929**					
IR	9.01					
P10	1					
P21	1					
TSI	271.8					
MVMPY	46.54					
ACCR	2.1					
K5	0.1					
Vehicle hours and accidents reduced						
Vehicle hours of delay reduced			**105902**	**per year**		
Accidents reduced			**9.1**	**per year**		

Fig. C.2 Inc mgt effectiveness potential example

Appendix D: Message Display Software for Southern State Parkway

This appendix describes the CMS semi-automatic control software employed in the freeway management system for the Southern State Parkway on Long Island in New York State [1].

The software performs the following functions:

- Provides CMS messages that display congestion conditions measured by system point detectors. Messages for several levels of congestion may be displayed.
- When congestion conditions are present on different sections of the roadway, the software prioritizes the messages to be displayed by location and severity condition. The priority scheme enables the limited messaging capability of the CMS to provide the most relevant information to the largest number of motorists.
- Provides the capability for the operator to alter the message.

Relationships Involving Geometry, Travel Time and Delay

The message generation technique is based on travel time and delay between certain key locations. Travel time is obtained from estimated speed from detector stations on the mainline of the freeway. A set of definitions and geometrical relationships to support the computation was defined and is discussed below.

As shown in Fig. D.1, a link represents a section of the mainline between vehicle access or egress points. The concept of a domain is introduced in order to relate data from freeway surveillance stations to mainline links. Domains relate links and CMS to the roadway locations receiving speed information from a particular detector station. Speed information from detector stations is filtered to remove short-term fluctuations. In some cases, the detector station may not physically be in the domain. As shown in the figure, each domain is assigned to a particular detector station. Domain boundaries are established at link nodes and at CMS. Where a link encompasses more than one detector station, domain boundaries are used to separate the regions for which each detector station will be employed.

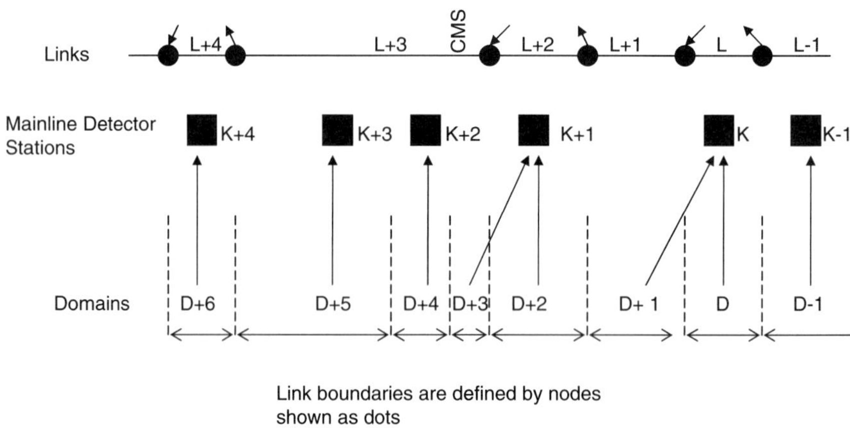

Fig. D.1 Link, domain and detector station relationships [1]

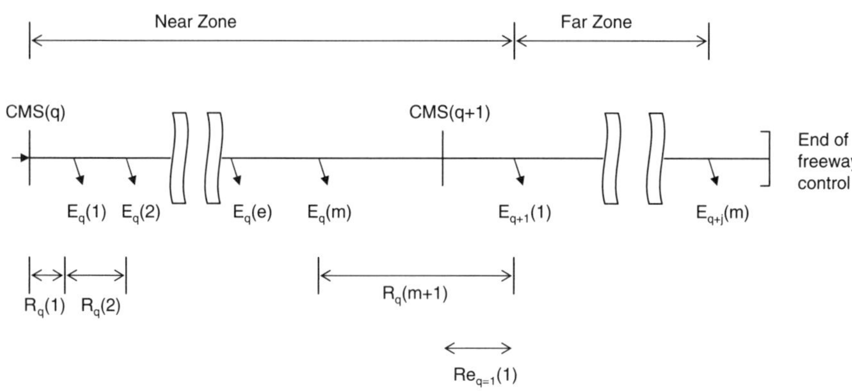

CMS	CMS locations
E	Exit ramps
R	Length of mainline roadway CMS control elements
q	Subscript for parameter definitions for near zone q
j	Subscript indicating end of far zone
m+1	Index for last element in near zone when there is at least one CMS in far zone
e	Index for element

Fig. D.2 Definitions for changeable message sign software [1]

Figure D.2 introduces the geometry and definitions used by the technique. In urban areas, CMS are often deployed at spatial intervals that encompass more than one exit ramp. Thus, a "near zone" is defined in Fig. D.2 as the distance from the CMS to the exit ramp just downstream of the subsequent CMS. This distance is serviced by one or more exit ramps $E_q(e)$. The "far zone," which is user defined, may encompass a number of subsequent CMS. A near zone and far zone are established for each CMS. As described later, the software develops different messages for the two zones.

Figure D.2 defines another data type, the "element" that is the distance between exit ramps. Each element includes one or more of the domains shown in Fig. D.1. Travel time for each domain is the quotient of domain length $R_q(e)$ and the speed provided by the detector station associated with each domain. Element travel time is the sum of the domain travel times within the element.

The system operator must establish a nominal travel time for each element. Element delay, the difference between element travel time and nominal travel time, is the basic component used for building the message.

Message Development

The following describes the message development capability of the software. The software provides two levels of congestion messages (delay and long delay) and a default message. For each element, the operator defines a delay threshold value that may trigger the message. Similarly, a long delay threshold value is also defined. A tentative delay indication or tentative long delay indication is declared for the element where prescribed thresholds are exceeded. The message is confirmed by the software if the declaration is valid for several computation intervals.

If the last element in the near zone (element m + 1 in Fig. D.2) exceeds the delay criterion for that zone, it is possible that the next downstream element is also congested. In this case, the software modifies the near zone boundary downstream to encompass those adjacent elements that experience delays.

Candidate Message Set Identification and Message Selection

Each candidate message is characterized by two factors:

- Intensity of delay (delay or long delay message classes)
- Freeway exits for which the delay condition applies

When the appropriate delay or long delay thresholds are exceeded, a set of candidate congestion messages is developed for each zone. Each candidate message identifies a contiguous group of elements experiencing congestion.

Because the CMS are limited in their message display capability, more candidate messages may be developed than can actually be displayed. Table D.1 identifies the priority classes for each CMS. The system operator may select the priorities. A commonly used priority order is a, b, c, d.

Table D.1 Message priority classes

	Near zone	Far zone
Long delay message set	a	c
Delay message set	b	d

Within each priority class, the messages are ordered geographically, the first message being closest to the motorist. CMS on the freeway mainline commonly display three or four lines of information. The software provides messages according to the priority selected within the CMS display constraints.

Reference

1. Southern State Parkway ITS early implementation project, Nassau County and Suffolk County (1997) New York State Department of Transportation

Appendix E: Washington State Fuzzy Logic Ramp Metering Algorithm

This description is based on the study by Taylor et al. [1], Neudorff et al. [2] and Klein [3]. Fuzzy logic has the ability to address multiple objectives (by weighing the rules that implement these objectives) and to implement the tuning process in a more user-friendly fashion (by the use of linguistic variables rather than numerical variables).

There are six inputs to the fuzzy logic controller (FLC). These include:

- Speed and occupancy from the mainline detector station located just upstream of the on-ramp merge
- Occupancy and speed from a downstream detector station. The station selected exhibits the maximum occupancy of selected downstream stations that have historically exhibited high flow breakdown rates
- Occupancy from a ramp queue detector typically located halfway between the ramp metering stop bar and the end of ramp storage
- Occupancy from the advanced queue occupancy detector at the upstream end of the ramp storage location

"Fuzzification" translates each numerical input into a set of fuzzy classes. For local occupancy and local speed, the fuzzy classes used are very small (VS), small (S), medium (M), big (B), and very big (VB). The degree of activation indicates how true that class is on a scale of 0 to 1. For example, if the local occupancy is 20%, the medium class would be true to a degree of 0.3, and the big class would be true to a degree of 0.8, while the remaining classes would be zero (Fig. E.1). The downstream occupancy only uses the very big class, which begins activating at 11%, and reaches full activation at 25% (Fig. E.2). The downstream speed uses the very small class, which begins activating at 64.4 km/h and reaches full activation at 88.5 km/h. The queue occupancy and advance queue occupancy use the very big class. For ramps with proper placement of ramp detectors, the parameter defaults are for activation to begin at 12%, and reach full activation at 30%. The dynamic range, distribution and shape of these fuzzy classes can be tuned for each input at each location.

The last step, called *defuzzification*, generates a single valued numerical metering rate based on the rule outcomes and the degree of activation.

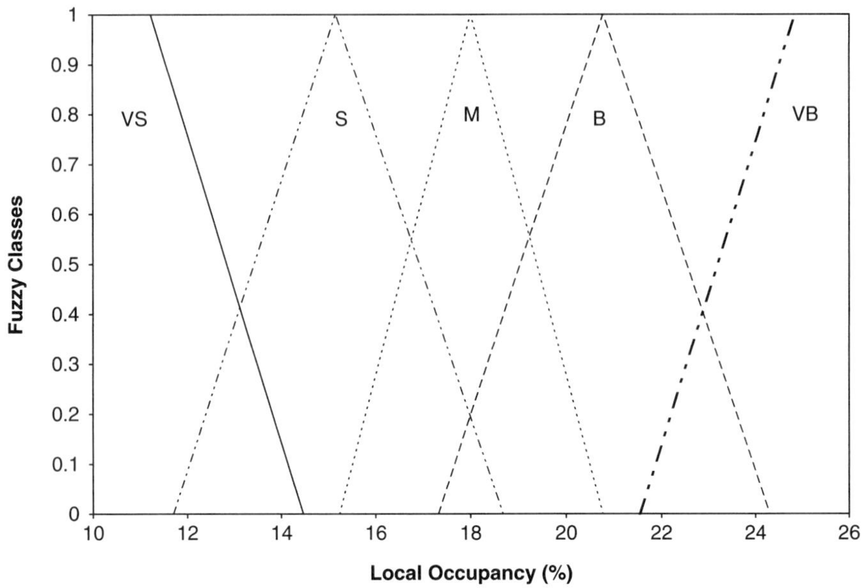

Fig. E.1 Fuzzy classes for local occupancy, redrawn. Presented at the 79th annual meeting of the Transportation Research Board, January 11, 2000, Washington, DC. Reproduced with permission of the Transportation Research Board

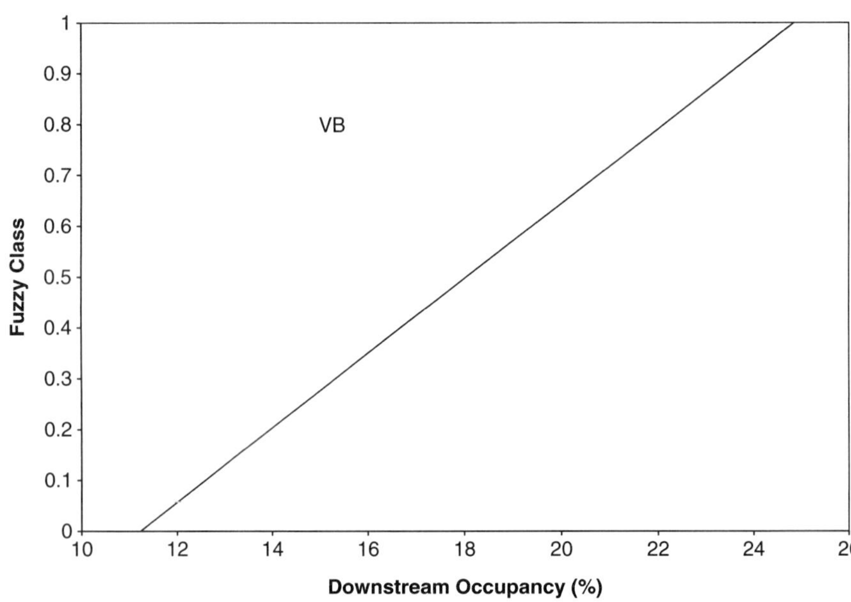

Fig. E.2 Fuzzy class for downstream occupancy [1], redrawn. Presented at the 79th annual meeting of the Transportation Research Board, January 11, 2000, Washington, DC. Reproduced with permission of the Transportation Research Board

Table E.1 Examples of fuzzy logic rules in the Washington state algorithm [1]

Rule	Default rule weight	Rule premise	Rule outcome
6	3.0	If local speed is VS AND local occupancy is VB	Metering rate is VS
10	4.0	If downstream speed is VS AND downstream occupancy is VB	Metering rate is VS
12	4.0	If advance queue occupancy is VB	Metering rate is VB

After the fuzzy states have been developed, weighted rules are then applied to develop the metering rate. Examples of weighted rules are shown in Table E.1.

References

1. Taylor C et al (2000) Results of the on-line implementation and testing of a fuzzy logic ramp metering algorithm. 79th annual meeting of the Transportation Research Board, Washington, DC
2. Neudorff LG et al (2003) Freeway management and operations handbook. Report FHWA-OP-04-003. Federal Highway Administration, Washington, DC
3. Klein LA (2001) Sensor technologies and data requirements fror ITS. Artech, Boston, MA

Appendix F: Benefits Model for Motorist Assistance Patrols

The Design ITS benefits model [1] for delay reduction for motorist assistance patrols is provided by the following expression:

$$TS = K12 \bullet K17 \bullet K35 \bullet K40 \bullet MTSSV \bullet LS$$

where

TS = Motorist time saved.

K12 = Annual service patrol stops per directional mile. The default value of 119 is based on INFORM ITS data and is based on an annual service patrol period of 2,600 h.

K17 = Fraction of stops for which assistance is provided. The default value of 0.46 is based on INFORM ITS data.

K35 = Correction factor. Design ITS provides the following default values:

K35 = 1.0 where peak hour level of service is D or worse.

K35 = 0.7 where peak hour level of service is C.

K35 = 0.0 where peak hour level of service is B or better.

K40 = Number of annual hours for which service is provided/2,600.

LS = Length of roadway segment.

MTSSV = Average delay reduced per assisted vehicle. The Design ITS default value is 95 hours [2]. It is based on a freeway with level of service D or worse.

References

1. Gordon RL (2007) Design ITS user's manual
2. Wohlschlaeger SD, Balke KN (1992) Incident response and clearance in the state of Texas: case studies of four motorist assistance patrols. Report No. FHWA/TX-92/1232-15. Texas Transportation Institute, College Station, TX

Index

A

AADT, 53–54, 72, 103, 104, 194
Accident, accident statistics, 16
ALINEA, 134–136
Alternate route(s), 4, 17, 20–21, 24, 67, 70–71, 82, 84, 86, 91–93, 98, 113, 116, 124–127, 142
Alternatives analysis, alternatives evaluation, 179–180
Archived data user service (ADUS), 181
Attribute, 28, 30, 147, 189
Automatic incident-detection (AID) algorithms, 65–66

B

Backbone (communications), 147, 150, 154
Bayesian inference, 75, 89
Benefits
- computation, 37
- and cost analysis, 6, 26, 37, 175–179
- and cost modeling, 37–38
- database, 37
- model, 15, 115, 143–144, 205
- economic benefit, 24–26, 180

Benefit-to-cost-ratio, 26
Binomial logit model, 87
Blank out signs, 67, 98
Bottleneck capacity, 18, 115, 127
Buffer time, 74

C

California Department of Transportation (CALTRANS), 2, 135, 137, 185
Capacity
- capacity reducing incidents, 81
- freeway capacity, 20, 54, 115, 119–126
- residual capacity, 52, 86
- roadway capacity, 43–45, 56–57, 94, 121, 134

Carbon dioxide, 26
Carbon monoxide, 26, 40, 78
Cellular telephone, 60, 66, 71, 82–83, 92, 149, 154–155
Center-to-center, 149, 150
Center-to-field, 151–156
Changeable message signs (CMS), 4, 5, 11, 22–24, 40, 82, 86, 89, 90, 93, 98–104, 110, 113, 133, 150, 154, 162, 166, 170, 177, 184–185, 197–200
Closed loop control
- closed loop manual control, 91–93
- closed loop semi-automatic control, 92–93

Code of Federal Regulations (CFR)
- 23CFR940, 7–8
- Part 940 of Title 23, 7, 24, 30

Cohort(s), 52–57, 183–184
Commercial radio, 82, 83, 110
Commercial vehicle delay, 26
Commercial vehicle inventory delay, 26
Commercial vehicles, 26, 35, 76, 78, 169
Communication(s)
- channel(s), 147–150, 155–156
- links, 44, 71, 157–158
- standards, 147–149

Compact disk (CD), 6, 183–185
COMPASS, 93
Computer aided dispatch (CAD), 166, 168
Concept of Operations (CONOPS), 8–10, 13, 90, 142, 161
Congestion
- non-recurrent congestion, 3–4, 15, 17, 20–24, 33–34, 37–38, 170
- recurrent congestion, 4, 14–15, 19–20, 23–24, 33–34, 37, 39, 46, 54, 81, 109–113, 127, 141, 162

Constraint(s), 2, 10, 12–13, 17, 30–32, 92, 97–98, 127, 136, 200
Construction, 11, 13–14, 34, 71, 81, 84, 98–99, 110, 162, 167, 173
Content factor, 87, 88
Contra-flow lanes, 107
Corridor, corridor delay reduction, 98, 143
Cost benefit analysis, 17, 26, 175–179
Costs, motorists' cost, 24

D

Data archiving, (data) mining, 163, 181
Data dictionaries, 149
Delay, 1, 3, 19, 22–28, 32, 40, 43, 45–47, 50–57, 61, 64, 74, 76–78, 81–82, 84, 86, 88, 92, 94–98, 105, 112–113, 116, 122–124, 126, 134, 140, 142–143, 163, 178, 180, 183–184, 193, 195, 197, 199
Demand-supply analysis, 127
Demand volume, 45, 46, 56, 60–61, 94, 96, 116
Density, 3, 17–19, 25, 63, 64, 92, 117, 119–121, 130, 135, 136
Design constraints, 31–32
Design ITS, 37–38, 40, 57, 74, 76, 78, 87, 89, 105, 143, 144, 176, 178, 187, 189, 205
Detector(s)
  acoustic detectors, 63, 171
  detector spacing, 63, 65, 75
  inductive loop detectors, 62, 131, 171, 181
  point (traffic) detector(s), 3, 18, 31, 59, 60, 62–66, 71–72, 90–93, 130, 150, 154, 159, 163, 167, 197
  probe traffic detection, 18
  probe traffic detector(s), 66 (AU: Probe traffic detectors is given as probe traffic detection in text)
  radar detectors, 62
  video processing detectors, 63
Distribution systems, 147, 149–150, 152
Diversion
  explicit diversion, 23, 84
  fraction
    probability(ies), 87–88
  messages, 23, 84, 86
  policy(ies), 84–86, 90, 95–97
  route, 14, 22, 33, 84, 97, 98
  strategy(ies), 87, 90–93
  volume, 86, 97, 98
Dudek, C.L., 22, 84, 99, 110
Dynamic traffic assignment, dynamic traffic routing, 93

E

E 511, 82, 83, 110
Effective capacity improvement, 122–123
Emergency evacuation(s), evacuation plan, 105–106
Emergency management services
  emergency service providers, 59, 70, 72, 161–162
  emergency services, 60, 161–162
Emergency vehicle(s), 3, 14, 71–72
Emergency vehicle turnarounds, 71
Emissions, vehicle emissions, 76, 78
Energy, 11
Environment(al), 8, 11, 12, 20, 25–27, 32, 99, 180
Equity, 27, 141–142
Evacuation route(s), 106–107
Evaluation
  prospective evaluation, 16, 24, 27, 175
  retrospective evaluation, 24, 27, 180
Evaluation Model for Freeway ITS Scoping, (EMFITS), 37–38, 187

F

False alarm, 66
Federal aid, 2, 7–8, 24, 30
Federal Highway Administration (FHWA), 2, 13, 15, 38, 72, 84, 187
Feedback, 16, 93, 105, 108, 170, 180
Fiber optic(s) (cable), 150–153, 155, 157
  multi-mode fiber, 151
  single mode fiber, 151, 152
Filter(ing), 132–134
Filter coefficient, 132, 133
Fire department, 71, 168
Flow
  breakdown, 119, 121–123, 130, 131, 134, 135, 201
  characteristics, 45, 119–126
  rate, 18, 19
  smoothing, 115, 123, 128
Free flow speed, 18
Freeway-to-freeway ramp metering, 140–141
Fuel consumption, 14, 24–26, 33, 76, 143, 180
Functional placement, CMS functional placement, 99–104
Fuzzy logic, 136, 201–203

G

Gini coefficient, 27, 142
Goals, 3, 10–16, 31
GPS, 66, 71, 82, 83, 89, 92, 110

Green signal split, 85, 97
Guide signs, 22

H
Hazardous materials, 73, 106, 149
High accident rate locations, 61, 191
Highway advisory radio (HAR), 40, 69, 82, 83, 89, 90, 110, 150, 154, 157, 159, 162, 167, 170, 177

I
IEEE 1512 standards, 69, 149
Incident
 clearance, 29, 30, 38, 45–47, 50, 51, 58–73, 94–96
 clearance time, 3, 15, 37, 50–52, 76
 command system, 59, 73
 delay, 46, 52–54, 61, 110
 detection, 59–67, 87, 94
 detection of an incident, 59–67, 87, 94
 detection time, 3, 34, 66
 dissemination model, 107
 duration, 47, 48, 51, 52, 56, 57, 74, 192
 frequency, 48
 incident-detection algorithm(s), 63–65
 incident-response plans, 58, 67
 information, 4, 38, 39, 58, 67, 81–108, 170
 Integrated Incident Management System (IIMS), 70
 Inter-agency communication, 67–70
 ITS standards, 69
 lane blocking incident, 46, 53–57
 management, 2, 3, 38, 44, 57–60, 67, 70, 149, 162, 165, 167, 168, 171, 193, 194
 management effectiveness, 65, 72–78, 184, 196
 recovery, 44, 58–72
 response, 25, 44, 58–72, 74, 149
 stages of an incident, 44, 58
Information level, 149
Integrated Incident Management System (IIMS), 70
IntelliDrive, 83
Intensity of deployment, deployment intensity, 32
Interchangeability, 2, 148, 149
Interoperability, 31, 34, 148, 149
ITS Deployment Analysis System (IDAS), 38, 187

J
Jam density, 120, 131–132

K
Kalman filter(s), 133

L
Lane closures, 50, 71, 81
Lane control signals, 1, 67, 107
Level of service (LOS), 19, 25, 32, 61, 75, 142–144, 177, 205
License plate readers, 66
Life cycle costs, 160, 176, 177
Linear programming, 127
Lorenz Curve, 27–28

M
Management concepts, 3–5, 29, 32–36
Marginal analysis, 26
Marginal values, 27
Media factor, 87–90, 105
Merge detector(s), 117–118
Message content, 20, 87, 88, 110
Message set(s), 90, 149, 199–200
Message strength, 20, 84, 87, 88
Metaline, 136
Metropolitan planning organizations (MPOs), 28
Microwave
 licensed microwave, 153, 157, 159
 unlicensed microwave, 150, 153, 156, 157
Mobility, 12, 14, 22, 25, 27, 180
Mode, 1, 19, 21, 81, 82, 151, 152, 162
Models, 3, 4, 6, 10, 15, 16, 20, 21, 37–40, 45–50, 52, 65, 71, 82, 86–87, 93, 100–104, 120, 165, 183, 187
Motorist information
 information to motorists, 5, 38, 39, 81–113, 162
 motorist messaging, 82–84
Motorist satisfaction, traveler satisfaction, 28, 29
Motorist service patrols, 15, 27, 37, 60, 67, 75, 194
Multi-attribute utility analysis, 28, 29
Multiplex, 150

N
National ambient air quality standard, 27
National ITS Architecture, 2, 7, 11, 31, 181

National Transportation Communications for ITS Protocol (NTCIP), 5, 69, 147–149, 167, 170
National Transportation Operations Coalition (NTOC), 73, 74
New York State Department of Transportation, 37, 49, 167
    New York State Department of Transportation Project Development Manual, 11 187
Node, 101, 102, 150, 157, 159, 197, 198
Non-diverting vehicle(s), 95–97

O

Oak Ridge Evaluation Modeling System (OREMS), 106
Objectives, 2–4, 7, 11–17, 22, 26, 27, 30, 32–36, 99, 127, 134, 176–178, 180, 201
Open loop control, 90–91
Open System Interconnection Architecture (OSI), 5, 147, 148
Operating procedure, 22, 23
Origin-destination, 24, 100–103, 184
Oxides of nitrogen (NOX), 14, 26, 40, 78

P

Passenger vehicle delay, 18, 21, 25, 26, 121
Peak hour, 27, 32, 36, 46, 61, 144, 194, 205
Peak period, 15, 27, 51, 52, 63, 84–86, 97, 112, 113, 120, 122–125, 130, 141
Peeta, S., 20, 87, 88
Percentage of objective satisfied, 27
Performance measures, 3, 13, 17, 24–30, 74
Police, 45, 59, 60, 63, 67, 69, 70, 75, 107, 161–163, 165, 168, 177, 194
Police patrols, 67
Policy, 22, 23, 81, 84, 86, 91, 92, 97, 110, 143, 162
Positive guidance, 22, 84
Presentation factor, 87, 88
Probability of incident detection, 66, 87
Probability theory, 75
Productivity, 12, 27
Project-level ITS architecture, 8
Propensity to divert, 86–88
Prospective evaluation(s), 16, 24, 27, 175
Public acceptance, 142–143
Public satisfaction, 25, 180
Public service access points (PSAPs), 60, 75, 194

Q

Quality of motorist information, 39, 40, 82, 105
Queue, 1, 5, 18, 44–46, 64, 73–76, 87, 94–97, 100, 115–127, 129, 135–138, 140–143, 161, 162, 184, 185, 201, 203
    detector(s), 58, 59, 72, 117, 118, 137, 140, 201
    discharge, 18, 119, 121–123

R

Ramp meter(s), 2, 27, 116–119, 134, 135, 137, 141–143, 171, 172, 185
Ramp metering
    isolated (ramp) metering, 5
    local (ramp) metering, 127–130, 135, 136
    metering rate(s), 115–119, 126, 127, 130, 132, 134–137, 140–142, 162, 201, 203
    metering strategies, 126–141, 163
    non-restrictive (ramp) metering, 5, 29, 30
    platoon metering, 117, 119, 123, 126, 135, 137, 144, 171, 1116
    pretimed (metering), 5, 117, 127–130, 137
    ramp metering benefits model, 143–144
    (ramp) metering plans, 163
    ramp meter installation, 117–119, 171
    restrictive (ramp) metering, 5, 24, 30, 115, 118, 123–128, 130–135, 137, 141, 144, 177
    system-wide ramp metering, 2, 5, 21, 113, 124, 127, 128, 134–135, 143
    traffic-responsive (metering), 5, 117, 127, 128, 130–134
Ramp storage, 137, 185, 201
Recurrent congestion, 3, 4, 14, 17, 19–20, 23–24, 33, 34, 37–39, 46, 54, 91, 109–113, 127, 141, 162, 170
Regional ITS Architecture, 6–8, 13, 161, 167, 169, 179
Remote traffic microwave sensor (RTMS), 153, 154
Requirements, 4, 7–9, 11–16, 30–32, 43, 50, 58, 62, 71, 74, 75, 78, 97, 100, 117–119, 126, 132, 135, 137, 140, 148–150, 157, 160, 163, 175, 180, 185
Restricted road use, 27
Retrospective evaluation(s), 24, 27, 180
Revealed preference survey, 88, 105
Road pricing, 27

S
Safety, 11–14, 25, 73, 107, 110, 123, 128, 135, 143, 149, 162, 168, 180
Satellite radio, 82, 83, 151
Saturated flow, oversaturated flow, 18
Scoping, 3, 10–13, 15, 24, 37–38, 100, 157, 187
Screening analysis for ITS (SCRITS), 38, 187
Secondary accidents, 1, 44–45, 76, 123, 163
Service rate, 115, 155
Set-point, 134–136
Shock wave(s), shock wave boundaries, 19
Shoulder accidents, 49, 54, 55
Shoulder periods, 109, 112, 113
Simulation, 16, 26, 29, 90, 106, 126, 127, 142, 163, 181
Smoothing, 115, 123, 128, 132
Special events, 14, 34, 81, 84, 109, 165
Speed
  average speed, 18, 25, 112, 117
  space-mean-speed, 18
  time-mean-speed, 18
Spillback, 123, 126, 129, 140
Stable flow, 119, 120
Stakeholder(s), 1, 2, 6–11, 13–15, 28, 34, 67, 97, 142, 165–167, 175, 179, 180
Standards, ITS standards, 8, 69, 147–149
Stated preference survey(s), 88, 105
Static sign(ing), 107
Strategic network management, 86–90
Strategy, 5, 9, 20, 27, 86, 88, 90–93, 107, 119, 126, 127, 134, 136, 143
Stratified zone metering, 136
Systems engineering, 2, 7–11, 13, 30, 161
  International Council on Systems Engineering (INCOSE), 8, 16
System Wide Adaptive Ramp Metering Algorithm, 136

T
Tail of queue, 58, 59, 72, 75, 87
Tandem metering, 119
Television traffic, 82
Throughput, 1, 2, 22, 24, 27, 116, 117, 123, 124, 143, 163
Time saved, 20–22, 54–57, 77, 97, 143, 205
Toll tag readers, 66, 92
Toll tags, 66
Traffic assignment, 20, 86, 87, 98
Traffic calming, 71, 72, 79
Traffic levels, 32, 36
Traffic signal(s), signal timing, 58, 59, 67, 90, 97, 106, 163, 173
Traffic signal preemption, 71
Trailblazers, 22
Transit Communications Interface Profiles, 149
Transmit, 66, 151
Transportation management center, 161–173
  traffic management center (TMC), 6, 22, 43, 50, 58, 60, 63, 67, 105, 107, 109, 126, 133, 141, 152, 155, 157, 159, 161, 171
Traveler behavior, 86
Travel time(s), 18–20, 23–27, 44, 45, 66, 70, 71, 84, 86, 92, 93, 110, 112, 123–126, 134, 141–143, 173, 197, 199
Travel time reliability, 25, 74, 112, 143
Travel time variation, 27, 112
Trip generation, 86

U
Undersaturated regime, 18
Unstable flow, 120
User satisfaction, 26, 27
Utility, 4, 6, 15, 17, 20, 28–30, 87, 88, 180

V
Vee diagram, 8, 11, 13, 180
Verification, 9, 13, 44, 58–60, 87
Volatile organic compounds, 26
Volume, 3, 14, 15, 18, 20, 23, 25, 45, 52–56, 59, 61, 63, 78, 85, 86, 94, 96, 101–103, 116, 117, 120–125, 127, 130–132, 134, 136, 137, 163, 178, 179, 181, 184, 194
Volume-to-capacity ratio, 15, 23, 25, 32, 46, 52–54, 113, 116, 184

W
Waiting time, 118, 136, 137, 141, 142
Wardrops's Principles, 23–24
  Wardrops's First Principle, 23
  Wardrops's Second Principle, 24
Weather, 6, 14, 20, 25, 33, 34, 63, 89, 98, 99, 106, 110, 134, 162, 165, 171
Wireless, 5, 71, 149, 150, 153–154, 156, 159, 168
Wireline, 5, 149, 151–153
Worksheet, 6, 54–57, 62, 76, 77, 95, 96, 103, 104, 125, 183–185, 192, 193

Z
Zone(s), 63, 130, 135, 136, 145, 180, 198–200

Springer